The Grand Strategies of Great Powers

W0081223

What is grand strategy and what is it good for? What are great powers, and which states are great powers today? What are the grand strategies available to great powers? What are the conditions under which a certain strategy is suitable and when should it be rejected? What are the factors affecting the success or failure of a given grand strategy? The present volume provides answers to these questions by introducing a typology of great power grand strategies, as strategies of rising, status quo, and declining powers, as well as through historical illustrations of each type. The reader is thus exposed to strategies such as divide and conquer, biding your time, opportunity strike, primacy, semi-detachment, concert, and appeasement through the experiences of leaders such as Bismarck, Peter the Great, Metternich, Deng Xiaoping, Neville Chamberlain, and Stalin. This analysis is then brought to bear on present developments in the grand strategies of the United States, China, and Russia. The volume should be of interest to both the academic and foreign policy-making communities, and in particular to students of international relations, diplomacy, history, and current international affairs.

Tudor A. Onea is Assistant Professor in the Department of International Relations of Bilkent University in Ankara, Turkey. He has been educated in Japan and Canada, has received his Ph.D. from Queen's University in Kingston, and has pursued post-doctoral stages in the United States, Canada, and Singapore. He has written a previous book, *US Foreign Policy in the Post-cold War: Restraint versus Assertiveness from George H.W. Bush to Barack Obama* (2013) and has published in *Review of International Studies*, *International Studies Review* (forthcoming), *European Journal of International Security*, and *International Relations*.

Routledge Studies in Modern History

For a full list of titles, please visit: https://www.routledge.com/history/series/MODHIST

The Grand Strategies of Great Powers

Tudor A. Onea

Routledge
Taylor & Francis Group

LONDON AND NEW YORK

First published 2021
by Routledge
2 Park Square, Milton Park, Abingdon, Oxon OX14 4RN

and by Routledge
605 Third Avenue, New York, NY 10017

First issued in paperback 2022

Routledge is an imprint of the Taylor & Francis Group, an informa business

Publisher's Note
The publisher has gone to great lengths to ensure the quality of this
reprint but points out that some imperfections in the original copies may
be apparent.

British Library Cataloguing-in-Publication Data
A catalogue record for this book is available from the British Library

Library of Congress Cataloging-in-Publication Data
Names: Onea, Tudor A., 1975- author.
Title: The Grand Strategies of Great Powers / Tudor A. Onea.
Description: London; New York, NY: Routledge/Taylor & Francis
Group, 2021. | Series: Routledge studies in modern history |
Includes bibliographical references and index.
Identifiers: LCCN 2020014240 (print) | LCCN 2020014241 (ebook) |
ISBN 9781138287181 (hardback) | ISBN 9781315268378 (ebook)
Subjects: LCSH: Strategy. | Great powers–History. | International
relations. | World politics.
Classification: LCC U162 .O527 2021 (print) | LCC U162 (ebook) |
DDC 327.1–dc23
LC record available at https://lccn.loc.gov/2020014240
LC ebook record available at https://lccn.loc.gov/2020014241

ISBN: 978-1-138-28718-1 (hbk)
ISBN: 978-0-367-53545-2 (pbk)
ISBN: 978-1-315-26837-8 (ebk)

DOI: 10.4324/9781315268378

Typeset in Times
by Deanta Global Publishing Services, Chennai, India

Contents

Acknowledgments

So let me now give praise to great women and men, and, in so doing, also to better scholars than I could ever hope to be, without whose help and guidance this book would have never happened. I would like in particular to express my gratitude for advice, mentorship, and friendship to Richard Ned Lebow, William Wohlforth, and Jack Levy, who got me interested in the field of grand strategy in the first place, and whose scholarship and whose example as serious scholars are and will always be an inspiration to me. Valuable suggestions and comments have been offered at different stages of this project, some even before it coagulated into an investigation of grand strategy, by Benjamin Valentino, Stephen Brooks, Deborah Larson, Janice Bially Mattern, Christopher Layne, John Schuessler, and Ted Hopf. I also want to take this opportunity to thank my colleagues and friends at Bilkent University for their constant encouragement and support, and for providing the best working environment I could have wished for. My utmost thanks to Berk Esen and Seckin Kostem, and also to Dimitris Tsarouhas, Eliza Gheorghe, Sam Hirst, Onur Isci, Ersel Aydanli, Caglar Kurc, and Tugba von Zu Bayar. I want to give particular credit and recognition to Erinc Yeldan, a stand-up person and outstanding Dean and friend. One cannot ever fully repay one's teachers, but in a small measure by paying it forward in one's own teaching. All the same, my thanks and appreciation should go to, in the order of my meeting them, Ion Mitran, Harald Kleinschmidt, and David Haglund.

Valuable research assistance has been offered for this project at Bilkent by a variety of students. I would like to thank Selim Yavuz, Selin Sahin, Yagmur Aytekin, Yusuf Yilmaz, Ahmet Sozen, Bilal Saglam, and Burak Yilmaz for their able efforts. I owe a very large debt of gratitude to my students at Bilkent: many, perhaps, the best insights of this book have arisen out of our courses and seminars.

There were times—many of them—when I thought that this second book of mine would never be written. As anyone who has faced the rough waters of the post-recession academic job market may attest, this is a contest of patience and motivation, which in my case has taken me around the world, literally. It is very tempting when things turn bad, as they often do, to lay down and just give up. I want to thank my family for helping me keep the true course, especially my formidable mother, Anda Maxim. Since no words would do justice for what you did for me, I hope you will accept the book itself as a way of saying thank you instead.

I know you appreciate books far more than journal articles: they have covers, so they must be serious scholarly endeavors weighed with authority.

Lastly, I want to thank the two strongest, kindest, and cleverest people I have ever encountered in my life: my grandparents, Nicolae Maxim and Edit Kadri Maxim. They made it through fascism, the Eastern Front, American bombings, Soviet occupation, expropriation for "unhealthy social origin," 1950s Stalinism, and the North Korean-like twenty-four-year police regime of Nicolae Ceausescu, without ever losing hope, or selling their souls. Thank you for teaching me the value of "meeting with triumph and disaster, and of treating these two impostors just the same."

1   Introduction

What is grand strategy?

More than 500 books on grand strategy saw print in the roughly three decades since the end of the Cold War. Proposing, debating, and criticizing grand strategy has become so commonplace that in 2011, a *Foreign Affairs* article observed that "whenever a foreign policy commentator articulates a new grand strategy, an angel gets its wings." Starting with the Goldwater-Nichols Act of 1986, which requires the US government to provide a yearly statement of its national security strategy, states have increased production of national strategies or white papers on defense or foreign policy. Presently, more than 30 countries have in place such a formal version of grand strategy. Departments of International Relations and Political Science around the world regularly teach courses on grand strategy, and the subject is even beginning to be offered as a major. Clearly, there is an on-going spike in research, policy-making, and educational interest in grand strategy.[1]

At the same time, there is an increasing need for order in how we think about grand strategy as a distinct field of study. The existing scholarship on grand strategy has pursued three main directions. The first is represented by single country studies of grand strategy, such as the grand strategy of the United States or of China, or historical examples such as the grand strategy of Philip II of Spain or of the Roman Empire. The second direction consists of whirlwind reviews of the evolution of grand strategy through history, bringing together pell-mell strategic theorists, such as Sun Tzu, Thucydides, Machiavelli, or Clausewitz, celebrated practitioners such as Bismarck, Churchill, and Kissinger, and the grand strategy of individual countries at given times, such as containment during the Cold War. Finally, the third direction views grand strategy as part of strategic behavior writ large. For instance, the magisterial volume by Sir Lawrence Freedman discusses, besides grand strategy, military strategy, the business strategy of companies, the strategy of social groups, and even theological strategy, such as the strategy that Milton's Satan resorts to against God.[2] The problem is that it is very hard to derive from either direction identifiable strategic behavior patterns that would yield general rules of how to analyze or select grand strategy. In other words, we are left with no answer to perhaps "the most basic question of social science: of what is this an instance?"[3] Indeed, we do not know if great powers have a grand strategy in place; if so, what sort of grand strategy are we dealing with; and if, and under what circumstances, a grand strategy that applies to one state would work for others.

By contrast, this book seeks to explore questions such as what are the grand strategies available to great powers. What are the conditions under which a certain strategy is suitable and when should it be rejected? What are the factors affecting the success or failure of a given grand strategy?

Definition

There are manifold definitions of grand strategy in the literature. A recent influential account by Nina Silove has attempted a classification of these definitions into three types. Accordingly, grand strategy may refer to a grand plan, to a grand organizing principle, or to grand behavior. In the first two types, grand strategy is formulated explicitly and intentionally by decision-makers; while in the latter, it takes shape and reveals itself progressively through the trials and tribulations of successive decision-makers. The difference between grand plan and grand principle comes down to the respective attention they give to detail. While the grand plan version of grand strategy considers in painstaking minutiae how exactly the grand strategy will be carried out, the grand principle version is content to trace a general outline, while leaving most of the specifics to be sorted out as they go. Meanwhile, the grand behavior version believes that not only the details of implementing the grand strategy, but also its very fundamentals, are worked out over time, as grand strategy evolves into a recognizable pattern. Despite their differences, all three versions agree that grand strategy is a long-term endeavor, that it guides the state in both war and peace, and that it seeks to advance the state's most important interests.[4] Silove concludes that "the most important reason for distinguishing the three concepts is that questions about the relationships among grand plans, grand principles, and grand behavior ought to be central to scholarship on grand strategy."[5]

This book opts for a definition of grand strategy as grand design or principle. But this design is not so sparse as Silove suggests, in the sense that grand strategy also incorporates aspects of the grand plan understanding of the concept, by considering the consequences, as well as the complications that are likely to arise from translating the guiding principle into practice. What grand strategy does not do, however, is work out every detail of how it is going to be executed in advance. Grand strategy is not decided through repeated meetings and debates between decision-makers huddled around a table, in an approximation of the ExCom in the Cuban missile crisis. Instead, grand strategy is *a great design, vision, idea, or master blueprint that orders and guides what a state does in interactions with the other actors, whether states or non-states, in the international system.* Grand strategy is thus the product of one or of several individuals' efforts at reflection concerning the present and likely future state of the international situation and of where one's state stands in it.

As such, grand strategy emerges fully shaped, like armored Minerva out of Jupiter's brow, rather than coming into being over time through trial, error, and improvisation, as held by the grand behavior interpretation.[6] Decision-makers use this grand vision or design as a guidebook for ulterior foreign policy, relying on conclusions already reached, instead of continuously starting the process from scratch. Foreign policy's job is therefore to translate grand strategy to individual country contexts, as well as to adjust it to developing circumstances, such as shifts in capabilities, the advent of new technologies and ideas, the availability of allies, or emergence of new enemies.[7]

Thus, while the strategic pattern that emerges over time necessarily differs in particular points from the original formulation, it is still recognizably guided by the same fundamentals, being handed over to the government from its predecessor, and being then passed on to its successor, as in a relay race. If the fundamentals change, then one deals with a different grand strategy. Consequently, the book is skeptical concerning the interpretation of grand strategy as "muddle through." Instead, it argues that the contribution made by decision-makers as they go, or by successive governments amounts to making corrections to an already laid down navigation course.

However, grand strategy is more akin to a navigation rutter, a naval pilot guidebook, than to a map, to which it is frequently compared. Unlike a map, a rutter was also concerned with offering directions pertaining to undertaking a voyage, such as describing channels, tides, harbors, and supply points, and warning of dangers such as reefs or enemy bases, and even providing advice on how to plot a course or how to treat disease on board. Correspondingly, grand strategy does far more than just picture where a state wants to go and how it should reach this destination. What it does is give instructions on which *goal* to pursue, *the path* to get there, and *the obstacles* to overcome. Thus, the chief intention of Frederick the Great, the King of Prussia, in his Political Testament of 1752, very much a statement of Prussia's grand strategy, was to "communicate to the posterity what I had learned by experience as a pilot who knows the stormy surroundings of the political seas; I undertake to indicate to them the reefs that they have to avoid and the harbors where they can find shelter." Likewise, the historian John Lewis Gaddis quotes the movie *Lincoln* (2012) for a similar insight:

> a compass will point you true north from where you are standing, but it's got not advice about the swamps and deserts and chasms that you'll encounter along the way. If in pursuit of your destination, you plunge ahead, heedless of obstacles, and achieve nothing more than to sink in a swamp … what's the use of knowing true north?[8]

Effectively, grand strategy can be understood as a trinity of fundamentals: *goals*, *means*, and *challenges*. A useful illustration is a "strategic triangle," in which each factor affects and is affected in turn by the other two (Figure 1.1).

States choose to pursue certain goals in the international arena by taking into account the power at their disposal and the ways to use it, as well the foreign opposition they will likely face.[9] They develop their own resources and choose tactics in light of the objectives they pursue, as well as of the magnitude of the

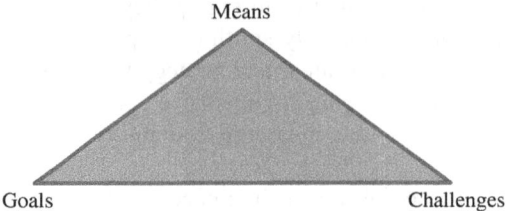

Figure 1.1 The strategic triangle.

resistance they anticipate. Finally, states identify which actors are most likely to oppose them depending on what they seek to achieve, and on their perception of how their resources and tactics fare by comparison. It follows that the strategic decision-making process does not occur in a chronological linear sequence, with one fundamental element predating and claiming paramount importance over the others. States do not pick their goals in a vacuum, based solely on their preferences derived from their peculiar cultural or ideological inclinations, or from the competing interests of domestic interest groups and coalitions—and only after that start to think about their resources or ways to use them or anticipate likely challenges. Nor do states simply do whatever they are able to do because of their means, while turning a blind eye to motives for action or to other states' reactions. And no grand strategy is defined just by the presence of an enemy, because for an enemy to exist and be identified as such, it should pose in the first place a threat to the state's goals, and its power and ways to use it compared to those of the state should be formidable enough to be a concern.

So how do states formulate their grand strategy? The hypothesis on which this book relies is that states choose to carry out a specific course of action according to their position in the international environment confronting them. Environment refers to all international phenomena to which the state's activities may be related.[10] Consequently, the international environment is much wider than what is commonly viewed as the international system. The environment includes not only systemic elements such as material capabilities, but also non-systemic ones: the non-material components of power, such as morale, competence, or diplomacy; geopolitical location and territorial layout; the history of interactions with other states and between third parties; the perceptions of decision-makers as to the interests, status, and capabilities of other states and of their own; technological innovation and diffusion; strategic culture (shared ideas about the character of their own state and of other polities, international relations, war, and peace); alignments; and, most importantly, the anticipation of other parties' likely response to one's action. The very essence of strategy according to Freedman consists in "interdependent decision-taking. The strategy of A depends on an assumption about the likely strategy of B, which in turn anticipates A's strategy and so on."[11] Therefore, grand strategy is the product of "environmental compulsion," in the sense that states identify goals, means, and challenges based on what sort of international strategic environment they face and anticipate.[12]

Grand strategy fundamentals

Grand strategy performs two functions vis-à-vis state goals. First, it helps define the national interest of the state. The goals that a state pursues beyond its borders range widely. Yet, in a last ratio, all states' goals concern possessions, material or non-material, to which they attach value, or in other words, that they see as valuable, whether this refers to territory, status, wealth, security, institutions, independence, way of life, or helping other human beings. States, and great powers especially, are not one-purpose organizations: they seek to achieve several of these values at once, in various bundles. Nonetheless, while states have manifold interests, hopes, wishes, and aspirations related to such things of value, these cannot be considered proper goals unless they warrant the expenditures of resources, that is to say incurring costs or sacrifices in order to attain them. Not everything has an equal value: the more valuable a thing is, the more cost or sacrifice it warrants. By implication, the most important goals, designated as vital interests or as national interest, are the ones that command the highest cost—the blood of its own citizens. Hence, these are the goals for which the state is willing to use armed force and suffer casualties.[13] The task of grand strategy is selecting which particular combination of goals deserves being pursued as the national interest of the state.

Second, grand strategy helps narrow down goals to one of three possible categories. They consist in acquiring or enhancing, in maintaining, or in minimizing or reversing the loss of valuable things. That is to say that states choose to either obtain additional value, preserve the value they already control, or seek to limit or revert the loss of value.[14] Consequently, one can distinguish between rising states, status quo states, and declining states, respectively. Grand strategy helps the state figure out in which category it belongs and which actions it should take as a result, since what does work for one type of state, may not work for the others.

Means refers both to power and to tactics. Power stands for the total amount of resources the state disposes of. As such, it designates a mix of geopolitical, demographic, military, economic, technological, political, social, and diplomatic components. Geopolitical power concerns the size and layout of the state's territory, its position on the map, and its natural resource endowment. Demographic power consists of the size of the population, as well as the age and the proportion of urban (hence likely more productive and better educated) population. Military power denotes the size, componence, equipment, and training of the armed forces. Economic power refers to wealth, comprising economic productivity measured through GDP, but also including human capital (health and education), infrastructure, and the natural environment. Technological power represents the state's ability to innovate, that is, to discover and develop, but also implement new technology, especially in the economic and the military fields. Political power is the competence of being able to use these various components in an efficient and cost-effective way. Social power, sometimes referred to as morale, and assessed through votes, polls, and public opinion, represents the ability to mobilize and to conserve public support in favor of a given course of action. Diplomatic power is the ability to obtain the support of other states, thus adding their power to your own, or to prevent them from joining the opponent's side.[15] Grand strategy selects which components to develop and employ, and in what proportion to best achieve the state's objectives.

Tactics refer to the ways and methods to utilize these resources. In the military field, tactics stand for a subcomponent of military strategy: while military strategy is about the overall planning of the war campaign, tactics are about combat, consisting in the disposition and maneuver of forces on the battlefield, so the former pertains to the general, while the latter to the field officers. In the context of international relations, the essential difference between tactics and grand strategy boils down to the fact that grand strategy pays attention to fundamentals, namely to the identification of ultimate goals and of challenges, matters that tactics ignore.[16] Examples of such tactics are resorting to war or coercion, concluding or exiting alliances, increasing or cutting down spending and taxation, increasing armaments or entering arms control agreements, extending or renouncing commitments, or engaging or breaking off negotiations. However, the same tactic can be used to achieve widely varying, and even contradictory ultimate objectives, whether defensive or aggressive. For instance, Barry Posen makes the case that the command of the commons, the exercise of predominant control by a dominant power such as the US over the seas, air space, outer space, and cyberspace, can be used *both* in the service of a grand strategy of primacy and of a grand strategy of restraint.[17] As such, tactics are essentially instruments used to meet immediate and intermediate objectives that help achieve greater ends. What grand strategy does is lay down a sequence of consecutive steps that lead to accomplishing these ultimate goals, and then chooses the most appropriate tactics for realizing each particular step.[18]

Finally, challenges refer to threats and obstacles. Threats concern the anticipation of impending harm by other actors, particularly by other states, to the things the state holds to be of value. It is important to underline that threats are not just about the value that the state currently owns, but also to new things of value that the state may seek to acquire or enhance, as well as to objects it is on the verge of losing or to value that it has owned once and seek to reacquire. The rule of thumb is that, no matter if the state is rising, status quo, or declining, it will have to contend with the opposition of other actors that seek to prevent it from achieving its national interest. This means that the grand strategy of a given state has to prevail against the active efforts of other states to prevent it from succeeding, since these rivals, in turn, seek to accomplish their own goals against the state's resistance. Therefore, for a state's grand strategy to work, it must prove superior to the adversary's own grand strategy. This requires not only accurately identifying one's likely opponents, but also anticipating what they are likely to do in response to one's own projected course of action, and devising a prepared response. A grand strategy must be interactive, or it is no strategy at all.[19]

Obstacles refer to the phenomenon the Prussian general Carl von Clausewitz, the foremost authority on military strategy over the last two centuries, labelled "friction." All states face inevitable limitations in their ability to control events while seeking to accomplish their goals. To make friction worse, these complications are cumulative, meaning, as Clausewitz put it, that "countless minor incidents the kind you can never really foresee" contribute to a less than optimal and steadily deteriorating state performance over the course of an endeavor. "Everything is simple in war," Clausewitz wrote, "but the simplest thing is difficult." The same goes for grand strategy.[20] Accordingly, for a grand strategy to be effective, it needs to take into account friction and devise solutions for minimizing its effects.

Some of the more common roots of friction involve complications of space, time, information, coordination, and misperception. States face a loss of strength gradient, meaning that their ability to project power diminishes the further away from their home base they seek to influence events. Rapidly unfolding developments, for instance during a crisis, force leaders to reach a speedy decision without the luxury of undergoing a careful deliberation of the alternatives open to them and under conditions of severe stress. The information available to decision-makers, particularly about the intentions of potential adversaries and allies, is limited, secret, deceptive, or/and unreliable. The more an action requires the cooperation of several parties, whether between distinct governmental bureaucracies, or between several states, the harder it is to ensure effective concerted action, due to miscommunication and miscoordination. Decision-makers exhibit psychological biases in correctly assessing incoming information on its own merits, rather than on how it fits in their pre-existing belief schemes, or in reconsidering decisions already taken in the face of fresh disconfirming evidence.[21]

A particularly important manifestation of friction that deserves a more extensive discussion consists in the absence or waning of domestic support, or, put differently, in the loss of morale. A grand strategy cannot succeed without extracting resources from the public in the form of taxes and military service—to this extent relevant domestic audiences need to support or refrain from opposing the line chosen by the government. In democracies, these audiences consist of voters: a government losing support will be simply replaced in elections. But even nondemocratic states face this problem, as they have to keep on their side key constituencies such as the army, the bureaucracy, the intelligence services, the clergy, or the party members in order to stay in office. This is to say that a grand strategy is not practical unless it is endorsed or at least tolerable to powerful domestic constituencies. As President Harry S. Truman put it, the best ideas in the world are of no benefit unless they can be carried out.[22] Thus, grand strategy resembles a game of chess played simultaneously on two chessboards—one domestic, the other international. Winning requires victory on both boards, which affect each other. While the first move takes place on the international board, taking into account the international environment, decision-makers cannot afford to neglect the state of play on the domestic board.[23] That is why grand strategy can be understood as a two-stage process. In the first stage of grand strategy formulation, states leave out any considerations of domestic preferences and interests, being driven only by the dictates of the international environment confronting them. But in the second stage of grand strategy selling, governments must also "sell" their preliminary design to relevant domestic audiences, whether bureaucrats, allied and rival political parties and interest groups, the media, or the general public. In the process of gaining their support or reducing their opposition through a mix of persuasion, manipulation, side payments, coercion, and compromise, the original design suffers inevitable modifications in regards to timing, extent, rhetoric, or/ and content.[24]

There is a basic rule to any good grand strategy: the fundamentals of goals, means, and challenges should be proportional to each other. If any of the three elements is disproportionate, the state is in danger of either doing too much or too little. To quote American political commentator Walter Lippmann: "the nation must maintain its objectives and its power in equilibrium, its purposes within its means and its means equal to its purposes, its commitments related to its resources and its resources adequate to its commitments."[25] If goals are too ambitious and exceed the state's resources, the state falls prey to imperial overstretch: its sum total of commitments is greater than its ability to sustain them simultaneously. As a consequence, the state is continually on call to address a never-ending series of emergencies without ever being able to muster sufficient power to address them all adequately, and bankrupting itself in the process. But if means overshadow goals, the state punches below its weight. It lacks the competence or/and morale to mobilize, organize, and employ its resources efficiently. This means passing on opportunities in favor of other states and competing as a result, at a disadvantage later on.[26] If challenges exceed goals, threats worsen over time, as the state ignores or refuses to address them, with eventual disastrous consequences. Conversely, if goals exceed challenges, the state will squander away resources without any pressing need of doing so, fighting essentially threats that are not there, resulting in a costly, suboptimal course of action. Meanwhile, if challenges exceed means, the state provokes enemies that are out of its league power-wise, separately or together, hence initiating a conflict it cannot hope to win. Finally, if means exceed challenges, the state is likely to engender unnecessary suspicion and enmity in other states, because without a manifest rival, there is no reason for preserving excessive capabilities other than for potential aggression.[27] In the words of the British military strategist Basil Liddell Hart, in regards to equilibrium between fundamentals, "an excess may be as harmful as a deficiency."[28]

Identifying grand strategy

How do we know a grand strategy when we see one? Indeed, "does grand strategy have to be articulated for it to be said to exist at all; and if not, can grand strategy be said to move a nation even when that nation's fluctuating roster of mostly incompetent leaders are unsure as to why they do anything?" As seen in the above, until recently, few states have bothered to issue an explicit statement of grand strategy through an appropriately labelled government document.

Furthermore, decision-makers frequently express the view that the state does not follow any grand strategy and may have no need for one, getting along on a case-by-case basis. For instance, President Barack Obama remarked that he needed neither a new grand strategy, nor a master strategist, such as George Kennan, who masterminded containment, to devise one for him. [29]

Grand strategy comes in many shapes and forms. It may be found in either private or public statements, such as, but not limited to, speeches, letters, memos, memoirs, treatises, editorials, books, journal articles, and political testaments. Its authors may be decision-makers, civil servants, diplomats, generals, scholars, and opinion leaders. Essentially, grand strategy understood as grand design is about ideas, which means that in order to assess whether a grand strategy exists or not, one has to glean if and how its logic affects the actions of the states. But ideas do not require being expressed in a single, explicit, official announcement of policy. For instance, although lacking any such formal declaration, the historian Geoffrey Parker argues that King Philip II of Spain's policies, propaganda, and secondary documents show consistent evidence of a unitary "global strategic vision," and, hence, constitute a grand strategy. Moreover, decision-makers may be actually unaware of the role strategic ideas play in their actions; or may be unable to trace the actual author or origin of the ideas they are busy putting into practice. Yet, this does not make them any less powerful. To paraphrase John Maynard Keynes' adagio, "practical men, who believe themselves to be quite exempt from any intellectual influence, are usually the slaves [in the sense of dutiful followers] of some defunct" academic scribbler or decision-maker from a few generations ago.[30]

Nevertheless, one should also be skeptical of the opposite view that "all states have a grand strategy whether they know it or not." Grand strategy does not come into place randomly, by accident, luck, or in a fit of absent-mindedness, regardless of what the state does, or of whether it means or not to produce a grand strategy. Instead, grand strategy is the result of a conscious, intentional effort at producing a guiding international principle. [31]

The risk with implying a grand strategy design solely from policy-makers' actions is that we may detect the presence of a grand strategy where in fact there is none, and then use our imaginary grand strategy to explain, predict, and then, in a self-fulfilling prophecy, recommend what line should the state follow.[32] Not every state necessarily has a grand strategy in place at all times. To avoid this risk and determine if a grand strategy is indeed present, it is therefore important to start by examining both the actions and statements of decision-makers towards a large number of other states for evidence of an organizing logic concerning the fundamentals of goals, means, and challenges. The next step would be to inquire whether this master blueprint has been *a constant factor in decision-making over time*, by looking for references and self-references to the existence of the grand design, actions that are to be expected if it is valid (counterfactuals), and counter-evidence from alternative theories that might explain the outcome.[33] If the actions of a state are not caused by a grand strategy, then one is likely to notice a considerable contradiction or oscillation between governments in fundamentals, meaning that the goals chosen, the resources and tactics used to pursue them, and the states considered as allies or enemies will shift considerably, particularly with every change in office; or will cancel each other, being at loggerheads.[34] The classic examples here are Imperial Germany prior to World War I, and Imperial Japan prior to World War II. Germany had initially pursued a grand strategy aimed at ensuring equality with Great Britain, based on the building of warships and the acquisition of colonies, but without fully abandoning it, it then went on to pursue a grand strategy aimed at achieving political and economic dominance in Eastern Europe. The result was that Germany pushed Great Britain into the camp of its continental foes (France and Russia) and ended up encircled. Japan pursued a concomitant grand strategy of land expansion in the north in China, and of sea expansion in the south, by seizing the colonies of Western powers in Southeast Asia. This left Japan overextended, while at the same time falling foul of the US. That is to say, that the absence of grand strategy is indicated by vacillation and self-contradiction. However, the absence of balance between strategic elements, goals to means, goals to challenges, and means to challenges, is not a sign of the state lacking a grand strategy, but rather of it pursuing a suboptimal strategy. To exemplify, taking on too many commitments such as the Soviet Union in the 1970s and 1980s, refusing to pursue objectives commensurate with the state's resources as Holland in the eighteenth century, and becoming caught in hostilities against too many powerful opponents as Nazi Germany in the 1940s are not instances of non-grand strategy, but simply of bad grand strategies. Therefore, even if a grand strategy is not balanced or coherent, it remains a grand strategy.[35]

Does grand strategy matter?

Grand strategy is more than a buzzword that stands in for foreign policy, military strategy, or any kind of goal-oriented behavior. Without grand strategy, we would be missing an essential part of decision-making that occurs at a higher level than in either foreign policy or military strategy. Furthermore, grand strategy refers to a type of decision-making distinct from either that of a CEO choosing the best way to compete against other firms, of a student contemplating revision for a final exam, or of a football team preparing for a big game.

Foreign policy refers to the policy of a state towards another state, i.e., how country A relates to country B. This is why one speaks of US policy towards China, or of Russo-American relations, but not of a grand strategy oriented solely towards a single country. Foreign policy is about managing state-to-state interactions, or, by extension, about the sum of such dyadic interactions. But foreign policy is not interested in how such interactions impact each other. This would make an analysis that restricts itself solely to the discussion of foreign policy, the equivalent of what the World War II US Chief of Staff and later Truman's Secretary of State George C. Marshall of the Marshall plan fame called "theateritis."[36] By this, Marshall meant the tendency of military commanders to think only in terms of the requirements of their independent theatres of operation, without considering how their intertwined actions affect the conduct of the entire war effort. This latter task falls to military strategy, or in the case of world politics, to grand strategy, which integrates separate state-to-state relations into a cohesive whole. Thus, assuming a system formed of four states, the grand strategy of state A will be concerned with how the behavior of state A towards state B is likely to affect its own relations with C and D, B's relations with C and D, as well as relations between C and D instead of thinking of relations with B, C, and D as autonomous. As the Cardinal de Richelieu suggested, a state should contemplate the effect of its actions on everyone, as well as the possible counter-reaction, and, as such, "negotiate without ceasing, openly or secretly, everywhere, even if it yields no immediate fruit and the expected one is not yet apparent." Hence, if the study of foreign policy is the study of decision-making or of how certain individual decision or constellations of such decisions are arrived at, the study of grand strategy is the study of how the various decision contexts are interdependent, in other words how they influence each other, as well as the decisions of the other states in the system. As Robert Jervis puts it, we are dealing with a system if two conditions are met: a) that a set of elements be interconnected so that changes in some elements or their relations produce changes in other elements or relations; and b) that the system exhibits properties or behaviors different from those of its constituent parts. Both conditions are met in the case of grand strategy: "outcomes cannot be understood by adding together the units or their relations." Therefore, while foreign policy is dyadic, grand strategy is essentially systemic.[37]

The second difference between grand strategy and foreign policy is the time-frame involved. Foreign policy is concerned with the management of relations with other states over the short- to mid-term, meaning the present and next foreseeable interval, ranging from a few months to a few years. By contrast, grand strategy requires long-term planning, lasting for years on end, often stretching on for decades, and sometimes being pursued for centuries, regardless of the various ideological, cultural, and personal preferences and backgrounds of the various leaders in charge. For instance, containment remained the guiding principle for US foreign policy throughout eight consecutive presidential administrations over five decades, even though each administration implemented it with its own adjustments. It is no accident that the Policy Planning Staff of the State Department states its mission as providing "a longer term, strategic view of global trends and frame recommendations for the Secretary of State to advance US interests and American values."[38]

One of the chief criticisms of grand strategy is that it ultimately represents a "standardless, incoherent concept whose popularity surge after the Cold War multiplied the lack of rigor with which it was employed." This is based on the grounds that grand strategy is usurping the term of strategy, which, by right, should have been confined to military studies. Critics also point out that the concept of grand strategy originated in military studies, before taking on a life of its own.[39] However, there is a substantial distinction between military strategy and grand strategy. The former deals with the structure of military forces, their deployment, and their use throughout a campaign. But military strategy is agnostic to questions of selecting the overall goals to be pursued, of developing non-military resources and of the tactics to utilize them, and of choosing whom to ally with or whom to oppose.[40] In a nutshell, military strategy is about achieving victory in war, but not about whether the war should be fought at all, against whom, to what extent, and for what purposes. All these tasks are the proper domain of grand strategy, which operates at the highest level of generality.[41] This makes military strategy one of several subdivisions of grand strategy, alongside economic, technological, social, or diplomatic strategies. Accordingly, grand strategy is holistic. It subsumes, overrides, and coordinates separate domain-specific strategies. As Edward Mead Earle wrote: "[grand] strategy ... is not merely a concept of wartime, but is an inherent element of statecraft at all times."[42]

Furthermore, although the term arose in military studies, grand strategy has come to reflect a distinct reality that in ages past was captured by the label raison d'état or reason of state. Raison d'état has a long and distinguished pedigree, having generated a vast body of literature from the seventeenth to the nineteenth centuries. It is about the rational determination and pursuit of a state's vital interests, in the sense that decision-makers need to choose the best available option considering the consequences likely to arise in a given situation. Accordingly, the choice should be made solely based on reason, leaving aside other considerations such as prejudices, grudges, biases, convictions, whether for or against a given course of action. Therefore, reason of state held that states had one "true policy," which anyone in their place, that is confronting the same array of circumstances, would have chosen. Essentially, this is the very same point made by "environmental compulsion": grand strategy is the solution dictated by the international environment a state is facing.[43] This means that grand strategy means something substantially different from military strategy. The two may be homonyms, but they are not synonyms.

Lastly, not every instance of goal-oriented behavior constitutes grand strategy, contradicting the view that "it's wrong to say that states have grand strategies, but people don't." Gaddis, for instance, argues that if grand strategy represents "the calculated relation of ends to means," then its principles should be "applicable to any endeavor in which means must be deployed in the pursuit of important ends." Freedman provides a similar argument, since strategy of any type requires a "less-than-straightforward path to reach destination or necessitate[s] a decision on resources and their effective application."[44] Nevertheless, grand strategy stands apart from the strategic behavior of corporations, doctors, students, political parties, sports teams, or individuals buying houses in two ways.

First, it assumes the presence of a contest or, as Clausewitz put it, "the collision of two living forces," in the sense of a party having to overcome an active opposition to reach its goals. The challenges to overcome are not just inanimate and involuntary. They consist of rival actors whose own objective is to prevent the actor from succeeding. This makes grand strategy similar to competing corporations, political parties, sports teams, or fighting armies, but different from everyday-life goal-oriented endeavors, such as choosing a career, deciding on a romantic partner, buying a house, making a presentation, developing a project, deciding to have surgery, or preparing for an exam. Second, grand strategy is, nevertheless, also distinct from other types of contests or competitions against other actors because of the magnitude of the stakes involved, which extend beyond individual life or death. As previously discussed, grand strategy selects vital goals for which laying down one's life is considered an acceptable sacrifice, and which involve the things that a society holds the most valuable. Thus, defeat in a grand strategic competition is unlike losing a sport contest, an election, or economic competition between corporations, where the circle of impacted people, in case of defeat, consists of the contestants themselves, and their narrow circle of supporters. Not even a military contest fully compares: a state may afford to lose a battle or even a war as long as the national interests of the state are safeguarded, for instance through concluding a satisfactory peace settlement, as was the case for France in the War of the Spanish Succession (1701–1714.) But if national interests, the values that a society holds most dear, are placed in jeopardy, every member of the community is going to suffer the impact. As Hal Brands writes, paraphrasing Trotsky: you may not care about grand strategy, but grand strategy still cares about you.[45]

Therefore, grand strategy is a legitimate subject of inquiry in its own right, as well as one of considerable, and likely, vital importance. However, what grand strategy is not is a panacea. Grand strategy does not help turn every given objective, no matter how farfetched, into reality. It is not a great equalizer that can be substituted for any disparity in power. And it is not an ultimate weapon that can bring down any foe, no matter how formidable, or smite any number of opponents. Grand strategy starts up by admitting, not by refusing to acknowledge, a state's weaknesses and limitations, what is possible and what is out of reach, and then proceeds to pick contests and endeavors that the state is likely to win. While it might not gain the upper hand in an unequal contest, it would know how to avoid being caught up in one. Furthermore, in contexts where other things are nearly equal, every factor that can confer an advantage, even a limited one over the competition, is of considerable significance. That is to say that the better strategy makes the difference between success and failure. Not even a state that is particularly powerful can ignore grand strategy, for instance by arguing that it is

> not nearly as important as grand strategists think, because countries tend to be judged by their actions, not their words. What really matters for great powers is power—national economic and military strength—and that speaks loudly and clearly by itself.[46]

Yet, power does not come with instructions on how to use it, which means that its inevitable misuse or abuse will, sooner or later, lead to setbacks at the hands of other greater powers or even savvier weaker opponents. True, a powerful state may afford to lose far more than other less well-endowed states, because it has larger resources to spare, but this does not mean that it can be insouciant to the point of throwing them out of the window. After all, how many successive losses can a powerful state afford before its power starts to dwindle significantly?

Accordingly, grand strategy matters because it lets decision-makers know what goals to pursue, under what circumstances, in relation to what states, and by what combination of resources and tactics. Thus, grand strategy brings order to a state's foreign undertakings and in so doing helps maximize efficiency. Without a grand strategy in place, resources would be wasted, challenges would remain unaddressed, objectives would be pursued haphazardly, and the state's actions would be self-defeating. A purely reactive, opportunist approach, made-up as events go, and adapted to each context, may work temporarily and in isolated settings, but will eventually fail as success in one area undermines the pursuit of objectives in others. Gaddis compares it to wandering blindfolded through a minefield. Sooner or later, one is guaranteed to step on a mine.[47]

Plan of the book

This book proceeds from the hypothesis that great powers—a category of states standing apart from the others on the basis of capabilities, behavior, and recognition—play the determinant role in how world politics is shaped at any given time. This does not mean that states that are not great powers are unimportant, or that they cannot pursue a grand strategy because they do not have the necessary resources to carry it out. Even if not playing on the board of global politics, a state is still involved in lower-level regional politics. Its rivals may not be the great powers, but instead its more even-matched neighbors. Hence, every state requires a grand strategy to guide and organize its relations with its peers, as well as with extra-regional great powers. Even relatively weak states' local interests can significantly impact the calculations of great powers. This was for instance the case in the Cold War, when the superpowers competed over the allegiance of states such as Egypt, Cuba, Vietnam, Ethiopia, or Angola.[48] But, in a last ratio, only great powers can affect the whole system by taking actions that produce consequences for each region. As such, they deserve particular scrutiny. It is no accident that in most International Relations courses taught around the world, apart from the discussion of one's own state, most examples are derived from the examination of great powers. They are the most impactful actors, like it or not. Consequently, this book focuses on discussing the grand strategies of great powers.

However, great powers do not choose from the same menu of grand strategies. There are three kinds of great powers: rising powers, status quo powers, and declining powers. Rising powers seek to gain a seat or to improve considerably their position at the great power table. Status quo powers seek to conserve and moderately improve their current position among great powers. Declining powers aim at a minimum to prevent and limit further losses, while at a maximum to recover what they have lost. The core argument of this book is that *rising, status-quo, and declining great powers follow different grand strategies*. Hence, great powers' grand strategies are classified into three basic categories: respectively, grand strategies of rising powers, grand strategies of status quo powers, and grand strategies of declining powers.

The rest of the book is organized as follows. Chapter Two lays down the criteria that make a state a great power, as well as the grounds of classifying a given great power as rising, status quo, or declining. The chapter also specifies which states have been and presently are great powers and their respective time tenure in the great power club. Finally, it places the great powers from 1648 to the present in categories as rising, status quo, and declining powers.

Meanwhile, Chapters Three to Six are about the grand strategies pertaining to each category. Each chapter offers a description of the tenets of the key grand strategies available to that respective great power type, illustrates each grand strategy through a historical case study, and advances a number of lessons for that respective power type derived from the case studies. Chapter Three is concerned with the grand strategies of rising great powers and examines self-strengthening, opportunity strike and fait accompli, divide and conquer, and reassurance and biding your time. It discusses respectively the cases of Peter the Great's Russia, Frederick the Great's Prussia, Otto von Bismarck's Germany, and Deng Xiaoping's China. Chapters Four and Five are concerned with the grand strategies of status quo powers: primacy, concert, semi-detachment, and containment. It has as case studies France from the Cardinal de Richelieu to Louis XIV, Clemens von Metternich's Austria, Canning's and Palmerston's Great Britain, and the US in the early Cold War. Chapter Six is devoted to the grand strategies of declining powers: lying in wait and appeasement. It considers respectively the cases of Stalin's Soviet Union, and of Great Britain from the turn of the century to the aftermath of World War II. Lastly, Chapter Seven concludes the book by providing a short discussion of the current grand strategy and strategic debates for the three present great powers of the first rank in the club: the US, China, and Russia; and by considering the implications for the often-neglected role of the strategist in world politics. For educators considering relying on this book for teaching grand strategy, a potential way of putting together a grand strategy course may be to start up with two sessions covering what is grand strategy as well as the definition, typology, and identification of great powers. The course would then continue with a weekly presentation of the individual strategies of the rising powers, status quo powers, and declining powers. It would then conclude with a three-week discussion of the grand strategies of the current great powers today, or/and in case one's state is not a great power, with an examination of its own grand strategy. Lectures can be profitably combined with seminar debates about the characteristics of each grand strategy, its application in the historical setting, and its possible application to today's great powers.

It is also important to mention a number of methodological disclaimers. This book is not meant as an exhaustive catalogue of grand strategies for each great power type. While alternative grand strategies are certainly conceivable, this discussion only aims to provide a useful starting point for future endeavors, by making available a basic classification scheme that could later on be added to, improved, revised, or expanded to cover non-great powers or even pre-modern polities. The book captures a limited, but reliable snapshot of great power grand strategy over the last four centuries, including the most salient strategic archetypes, strategists, and influential great powers. Interested educators should, by all means, add their own grand strategies of choice, use different labels where they think they may be appropriate, or substitute different case studies.

It is also possible to conceive of alternative schemes for classifying the grand strategies of great powers, although it should be noted that there have been remarkably few such efforts previously.[49] The existing ones focus on either the role of geopolitical factors (naval vs. land strategies), or on grand strategies as different forms of the balance of power. This book considers that both these approaches suffer from inherent limitations in terms of providing a wide-breadth typology of the various forms of great power grand strategy. The naval vs. land classification has little to say about the varied grand strategies available to land powers, which constitute by far the majority of great powers. Besides, the naval strategy may actually be about a particular instance of grand strategy, discussed by this book as semi-detachment, and specific to Great Britain. Meanwhile, the balancing typology is foremost concerned with status quo powers, being less relevant for the grand strategies of declining and rising powers. Furthermore, balancing may actually be about tactics, such as the formation of a grand coalition, bandwagoning, and buck-passing, not about grand strategy; and, again, may be concerned with a specific grand strategy, discussed here as containment.

The book is also aware of the risk of selection bias in choosing cases that are highly likely to confirm the grand strategy hypothesis being offered for each. However, the book makes no claim to seek to explain in-depth the puzzle of why the respective great power took that particular course of action. In other words, it relies more on narrative description than on hypothesis testing on the case studies. There are two reasons for this way of proceeding. First, this is the entrenched trend in the existing literature.[50] Second, and more important, the avowed purpose of the book is to catalogue grand strategies and showcase what their abstract main points look like when translated into practice. Hence, providing historical narratives for the case studies is an acceptable trade-off at this point in order to achieve this purpose. However, supplementary testing for the specific cases presented, as well as for the validity of the grand strategy hypotheses in further case studies, is both advised and welcome. An objection may also be raised about the way the presentation of each individual grand strategy is organized: should not one start instead with the case study, and then proceed to extract from it a number of lessons to be distilled into a grand strategy? Yet, this manner of proceeding also comes with a sizable practical drawback from a teaching perspective: the student/reader has to go through the case study twice, once before being made aware of the actual grand strategy, and once after, so as to see how the grand strategy actually worked in practice. To avoid such redundancy, the book prefers to present first the main tenets of the grand strategy, which then can be traced throughout the case study.

Notes

1 Silove, "Beyond the Buzzword," 27–8. From 1990 to 2018, the exact number is 518 books published on the subject. Library of Congress Online Catalog at https://catalog.loc.gov/vwebv/search? For the angel gets its wings comment see Drezner, "Does Obama Have a Grand Strategy," 61. For examples of formal grand strategy documents

see inter alia White House, "National Security Strategy of the United States, December 2017," at https://www.whitehouse.gov/wp-content/uploads/2017/12/NSS-Final-12-18-2017-0905.pdf; "Russian Federation National Security Strategy, December 2015" at http://www.ieee.es/Galerias/fichero/OtrasPublicaciones/Internacional/2016/Russian-National-Security-Strategy-31Dec2015.pdf; "China's Military Strategy White Paper, May 2017," at http://english.gov.cn/archive/white_paper/2015/05/27/content_2814 75115610833.htm; "Strategic Review of Defense and National Security 2017," at www.defense.gouv.fr/content/ download/514655/ 8664340/file/ Revue+strategique+de+defense+et+de+securite+nationale+2017.pdf; HM Government, "National Security Strategy and Strategic Defense and Security Review 2015," at https://assets.publishing.service.gov.uk/government/uploads/system/uploads/attachment_data/file/555607/20 15_Strategic_Defence_and_Security_Review.pdf. For academic programs in grand strategy see the Brady Johnson Program in Grand Strategy at Yale University at https:// grandstrategy.yale.edu/, the American Grand Strategy program at Duke University at https://sites.duke.edu/agsp/, and the Master of Science in International Strategy and Diplomacy at London School of Economics at http://www.lse.ac.uk/study-at-lse/G raduate/Degree-programmes-2018/MSc-International-Strategy-and-Diplomacy.

2 For individual countries' grand strategies see inter alia on the US, Art, *Grand Strategy for America*; Dueck, *Reluctant Crusaders*; Brands, *What Good Is Grand Strategy?*; Posen, *Restraint*; on China, see Swaine and Tellis, *Interpreting China's Grand Strategy*; Goldstein, *Rising to the Challenge*; on Japan, Samuels, *Securing Japan*; Paine, *Japanese Empire*; on Russia, LeDonne, *Grand Strategy of Russian Empire*; Luttwak, *Grand Strategy of Soviet Union*; on India, Bajpai, Basit, and Krishnappa, eds. *India's Grand Strategy*; on more remote historical cases, see Luttwak, *Grand Strategy of Roman Empire*; Luttwak *Grand Strategy of Byzantine Empire*; Parker, *Grand Strategy of Philip II*. For grand strategy throughout history, see Kennedy, *Strategy and Diplomacy*; Kennedy, ed., *Grand Strategies in War and Peace*; Murray, Knox, and Bernstein, eds. *Making of Strategy*; Murray, Sinnreich, and Lacey, eds. *Shaping of Grand Strategy*; Balzacq, Dombrovski, and Reich, eds., *Comparative Grand Strategy*; Gaddis, *On Grand Strategy*. On strategy in general see Gray, *Modern Strategy*; Hill, *Grand Strategies*; Freedman, *Strategy*; Milevski, *Evolution of Grand Strategic Thought*.

3 Baldwin, "Concept of Security," 6.

4 Silove, "Beyond the Buzzword." For some of the more influential discussions of grand strategy, see Earle, "Introduction," vii–xi; Liddell Hart, *Strategy*; Kennedy, *Grand Strategies in War and Peace*; Craig and Gilbert, "Reflections on Strategy in the Present and Future," 863–71; Brands, *What Good Is Grand Strategy*; Martel, *Grand Strategy in Theory and Practice*.

5 Silove, "Beyond the Buzzword," 55; also see, for the lack of generalizability, Thierry Balzacq, Peter Dombrowski, and Simon Reich, "Is Grand Strategy a Research Program? A Review Essay," *Security Studies*, published online October 2018 at https://www.tan dfonline.com/doi/full/10.1080/09636412.2018.1508631, 23.

6 Mitzen, "Illusion or Intention?" For the muddle through perspective see Popescu, *Emergent Strategy and Grand Strategy*; Trachtenberg, "Making Grand Strategy"; Edelstein and Krebs, "Delusions of Grand Strategy"; Reich and Dombrowski, *End of Grand Strategy*.

7 Brands, *What Good Is Grand Strategy*, 5.

8 Friederich des Grossen, "Die Politischen Testamente," at https://archive.org/stream/ diepolitischente00freduoft/diepolitischente00freduoft_djvu.txt ,1; Gaddis, *On Grand Strategy*, 17; also see for a similar insight Deibel, *Foreign Affairs Strategy*, 4.

9 Leffler, *Preponderance of Power*, ix; Layne, "From Preponderance to Offshore Balancing," 88.

10 Sprout and Sprout, "Environmental Factors," 316.

11 Freedman, *Deterrence*, 43–4.
12 For the concept of environmental compulsion, see Baron and Parent, "Elder Abuse," 201.
13 Wolfers, *Discord and Collaboration*, 67–80, 71; Gilpin, *War and Change*, 19–25.
14 Wolfers supports a somewhat similar division, but he distinguishes between possession goals—the enhancement or preservation of value—and milieu goals, which seek to shape conditions beyond one's borders; also, he does not include restricting or recuperating the loss of value. Wolfers, *Discord and Collaboration*, 73–4.
15 Tellis, Bially, Layne, and McPherson, *Measuring National Power*; Morgenthau, *Politics Among Nations*.
16 The differences between tactics and strategy are best discussed in Freedman, "The Meaning of Strategy, Part I"; and Lawrence Freedman, "The Meaning of Strategy, Part II"; also see Clausewitz, *On War*, Book II, 128–32; Earle, "Introduction," viii; Liddell Hart, *Strategy*, 337–8; Martel, *Grand Strategy*, 27–30; Luttwak, *Strategy*, chaps, 5–10.
17 Posen, *Restraint*, XII–XIII; Posen, "Command of the Commons," 7–9, 21; MacDonald and Parent, "Graceful Decline?"
18 Deibel, *Foreign Affairs Strategy*, 24–32.
19 Cohen, *Threat Perception*, 4–6; Baldwin, "Concept of Security," 11, 13; Freedman, *Strategy*, xi–xii.
20 Clausewitz, *On War*, Book I, 7; Gaddis, *On Grand Strategy*, 24.
21 Boulding, *Conflict and Defense*, 260–2, 268–9; Lebow, *Between Peace and War*; Jervis, *Logic of Images*; Mearsheimer, *Why Leaders Lie*; Allison and Zelikow, *Essence of Decision*; Jervis, *Perception and Misperception*; Friedberg, *Weary Titan*.
22 Truman quoted in Larson, *Origins of Containment*, 129; Christensen, *Useful Adversaries*; Ripsman, Lobell, and Taliaferro, *Neoclassical Realist Theory*.
23 Putnam, "Diplomacy and Domestic Politics"; Dueck, *Reluctant Crusaders*; Ripsman, Lobell, and Taliaferro, eds. *Neoclassical Realism, State, and Foreign Policy*; Rosecrance and Stein, eds. *Domestic Bases of Grand Strategy*.
24 William Martel similarly argues for a two-step grand strategy process consisting of articulation and implementation. The distinctions consist in that by articulation, he refers solely to the state selecting and prioritizing national goals; and that articulation and implementation do not occur at the international and domestic levels as in this book's argument. Martel, *Grand Strategy*, 37–56; also see Deibel, *Foreign Affairs Strategy*, 11–2.
25 Lippmann, *US Foreign Policy*.
26 Kennedy, *Rise and Fall of Great Powers*. Considerably less exists on under-stretch or states failing to develop commitments to match their resources. For possible illustrations see Zakaria, *From Wealth to Power*; Israel, *Dutch Republic*.
27 Jervis, *Perception and Misperception*, chap. 3; Jervis, "Perceiving and Coping with Threat," 16–7.
28 Liddell Hart, *Strategy*, 336.
29 McDougall, "Can the United States Do Grand Strategy?"; Balzacq, Dombrovski, and Reich, eds., *Comparative Grand Strategy*, 3, 1–24; Remnick, "Going the Distance."
30 Parker, *Grand Strategy of Philip II*, 2–9; Brands, *What Good Is Grand Strategy*, 5–6; Silove, "Beyond the Buzzword," 40–2; Keynes, *General Theory*, 383–4.
31 For the idea that every state inherently has a grand strategy, see Luttwak, *Grand Strategy of the Byzantine Empire*, 409; also see Silove, "Beyond the Buzzword," 49–50.
32 See for the warning that the influence of ideas risks being ubiquitous Kowert and Legro, "Norms, Identity, and Their Limits," 451–97. For a method allowing one to trace the influence of ideas into policy through the stages of emergence, propagation (cascade), and internalization, see Finnemore and Sikkink, "International Norms Dynamics and Political Change," 895–905.
33 Silove, "Beyond the Buzzword," 50.

34 Silove argues that the absence of grand strategy understood as a grand principle would be characterized by the absence of long-term goals or by the failure to identify means. Ibid., 50. But how can this be attested in practice particularly without an explicit single statement? The behavior of the state would have to be erratic, showing no continuity over time, or would be made-up on the spot without an attempt to coordinate between policies directed at various countries and regions. Other tell-tale signs of the absence of grand strategy are prostration, hypervigilance, and panic, as the government in charge abdicates decision by doing what everyone else is doing or succumbing to fight-or-flight impulses. Janis and Mann, *Decision-Making*, 59–62; Lebow, *Between Peace and War*, 119–47.

35 Martel, *Grand Strategy*, 32; Silove, "Beyond the Buzzword," 48–9.

36 John Lewis Gaddis, "What Is Grand Strategy?" Karl Von Der Heyden Distinguished Lecture, Duke University, February 26, 2009, available at http://indianstrategickno wledgeonline.com/web/grandstrategypaper.pdf.

37 Jervis, *System Effects*, 6; Brands, *What Good Is Grand Strategy*, 3; Deibel, *Foreign Affairs Strategy*, 9, 21–2; Layton, "Idea of Grand Strategy"; Elliott, *Richelieu and Olivares*, 129. For descriptions of foreign policy as a field, see Carlsnaes, "Foreign Policy"; Hudson, "Foreign Policy Analysis." For the view that grand strategy, understood as the theory of how a state ensures its military security, represents only a branch of foreign policy, which may have non-military goals such as freedom or prosperity, see Posen, *Restraint*, 1–2, 4; Dueck, *Reluctant Crusaders*, 11.

38 Kennedy, "Grand Strategy in War and Peace: Towards a Broader Definition," in Kennedy, ed., *Grand Strategies*; Ripsman, Lobell, and Taliaferro, *Neoclassical Realist Theory*, 83–4; Deibel, *Foreign Affairs Strategy*, 14–6.

39 Milevski, *Evolution of Grand Strategic Thought*, esp. 141; for the origins of the term grand strategy, see Freedman, "Meaning of Strategy, Part I." It should also be added that several scholars conflate grand strategy and military strategy. For instance, Posen writes that "grand strategy is ultimately about fighting, a costly and bloody business" and more to the point just about security (grand strategy is "a state's theory about how it can best 'cause' security for itself"); while Robert Art understands grand strategy as telling the nation's leaders what goals to aim for and how to best use military instruments to attain them. Posen, *Restraint*, 1; Posen, *Sources of Military Doctrine*, 13; Art, *Grand Strategy for America*, 1–2.

40 This is precisely the point of Clausewitz's maxim that war represents the "continuation of political intercourse, carried out by other means." Left to its own devices, war tends to become total war, pure destruction. It is therefore unleashed for political objectives, once other means such as diplomacy and economic pressure have proved ineffective. But it also should end once the political objective is met, instead of being pursued to its logical end, the complete annihilation of the enemy's population. Thus, for Clausewitz, unlike for von Moltke the Elder, the ultimate authority in charge remains the government, not the military commander. Clausewitz, *On War*, Book 8, 6.

41 Kennedy, *Grand Strategy*, 4–5; Craig and Gilbert, "Reflections," 869; Gaddis, "Containment," 28; Martel, *Grand Strategy*, 31.

42 Earle, "Introduction," VIII; Deibel, *Foreign Affairs Strategy*, 3–10, 13–4.

43 Craig and Gilbert, "Reflections on Strategy," 869; Meinecke, *Machiavellism*, 1; Gilbert, *To the Farewell Address*, 95–104; Church, *Richelieu and Reason of State*.

44 Gaddis, "What Is Grand Strategy"; Gaddis, *On Grand Strategy*, 19–21; Freedman, *Strategy*, IX–X.

45 Clausewitz, *On War*, Book I, 1; Brands, *What Good Is Grand Strategy*, 6.

46 Drezner, "Does Obama Have a Grand Strategy?"; also see Yetiv, *Absence of Grand Strategy*; Popescu, *Emergent Strategy*; Edelstein and Krebs, "Delusions of Grandeur."

47 Gaddis, "Containment and the Logic of Strategy"; for a comprehensive examination and response to critiques of strategy see Betts, "Is Strategy an Illusion?"

48 Stephen Krasner, "An Orienting Principle for Foreign Policy," *Policy Review*, October 1, 2010, available at https://www.hoover.org/research/orienting-principle-foreign-polic y; Silove, "Beyond the Buzzword," 51; Balzacq, Dombrovski, and Reich, "Is Grand Strategy a Research Program," 28.
49 See esp. Mearsheimer, *Tragedy*, 139; Schweller, *Deadly Imbalances*, 66–75; Schweller, "Managing the Rise of Great Powers: History and Theory," 7–17.
50 Balzacq, Dombrovski, and Reich, "Is Grand Strategy a Research Program," 26–7.

Bibliography

Allison, Graham and Philip Zelikow. *Essence of Decision: Explaining the Cuban Missile Crisis*. New York: Addison Wesley, 1999.
Art, Robert. *A Grand Strategy for America*. Ithaca: Cornell University Press, 2003.
Bajpai, K, S Basit, and V Krishnappa, eds. *India's Grand Strategy: History, Theory and Cases*. New Delhi: Routledge India, 2014.
Baldwin, David. "The Concept of Security." *Review of International Studies* 23 (1997): 5–26.
Balzacq, Thierry, Peter Dombrovski, and Simon Reich, eds. *Comparative Grand Strategy: A Framework and Cases*. Oxford: Oxford University Press, 2019.
Baron, Joshua and Joseph Parent. "Elder Abuse: How the Moderns Mistreat Classical Realism." *International Studies Review* 13, no. 2 (2011): 193–213.
Betts, Richard. "Is Strategy an Illusion?" *International Security* 25 (Fall 2000): 5–50.
Boulding, Kenneth. *Conflict and Defense: A General Theory*. New York: Harper, 1962.
Brands, Hal. *What Good Is Grand Strategy? Power and Purpose in American Statecraft from Harry S. Truman to George W. Bush*. Ithaca: Cornell University Press, 2014.
Carlsnaes, Walter. "Foreign Policy." In *Handbook of International Relations*, edited by Walter Carlsnaes, Thomas Risse, and Beth Simmons. Los Angeles: SAGE, 2013.
Christensen, Thomas. *Useful Adversaries: Grand Strategy, Domestic Mobilization, and Sino-American Conflict, 1947–1958*. Princeton: Princeton University Press, 1996.
Church, William. *Richelieu and Reason of State*. Princeton: Princeton University Press, 2015.
Clausewitz, Carl von. *On War*, Peter Paret trans. Princeton: Princeton University Press, 1976.
Cohen, Raymond. *Threat Perception in International Crisis*. Maddison: University of Wisconsin Press, 1979.
Craig, Gordon and Felix Gilbert. "Reflections on Strategy in the Present and Future." In *Makers of Modern Strategy: From Machiavelli to the Nuclear Age*, edited by Peter Paret, Gordon Craig, and Felix Gilbert. Princeton: Princeton University Press, 1986.
Deibel, Terry. *Foreign Affairs Strategy: Logic for American Statecraft*. Cambridge: Cambridge University Press, 2007.
Drezner, Daniel. "Does Obama Have a Grand Strategy: Why We Need Doctrines in Uncertain Times." *Foreign Affairs* 90 (July 2011): 57–68.
Dueck, Colin. *Reluctant Crusaders: Power, Culture, and Change in American Grand Strategy*. Princeton: Princeton University Press, 2006.
Earle, Edward Mead. "Introduction." In *Makers of Modern Strategy: Military Thought from Machiavelli to Hitler*, edited by Edward Mead Earle, Gordon Craig, and Felix Gilbert. Princeton: Princeton University Press, 1943.
Edelstein, David and Ronald Krebs. "Delusions of Grand Strategy: The Problem with Washington's Planning Obsession." *Foreign Affairs* 94 (November/December 2015): 109–16.

Elliott, JH. *Richelieu and Olivares*. New York: Cambridge University Press, 1991.

Finnemore, Martha and Kathryn Sikkink. "International Norms Dynamics and Political Change." *International Organization* 52 (September 1998): 887–917.

Freedman, Lawrence. *Deterrence*. Malden: Polity, 2004.

Freedman, Lawrence. *Strategy*. Oxford: Oxford University Press, 2013.

Freedman, Lawrence. "The Meaning of Strategy, Part I: The Origins." *Texas National Security Review* I (November 2017): 90–105.

Freedman, Lawrence. "The Meaning of Strategy, Part II: The Objectives." *Texas National Security Review* I (March 2018): 34–56.

Friedberg, Aaron. *Weary Titan: Britain and the Experience of Relative Decline, 1895–1905*. Princeton: Princeton University Press, 1988.

Gaddis, John Lewis. "Containment and the Logic of Strategy." *National Interest* 10 (Winter 1987): 27–38.

Gaddis, John Lewis. *On Grand Strategy*. New York: Penguin Press, 2018.

Gilbert, Felix. *To the Farewell Address: Ideas of Early American Foreign Policy*. Princeton: Princeton University Press, 1961.

Gilpin, Robert. *War and Change in World Politics*. New York: Cambridge University Press, 1981.

Goldstein, Avery. *Rising to the Challenge: China's Grand Strategy and International Security*. Stanford: Stanford University Press, 2005.

Gray, Colin. *Modern Strategy*. Oxford: Oxford University Press, 1999.

Hill, Charles. *Grand Strategies: Literature, Statecraft and World Order*. New Haven: Yale University Press, 2010.

Hudson, Valerie. "Foreign Policy Analysis: Actor-Specific Theory and the Ground of International Relations." *Foreign Policy Analysis* 1 (March 2005): 1–30.

Israel, Jonathan. *The Dutch Republic: Its Rise, Greatness, and Fall, 1477–1806*. Oxford: Oxford University Press, 1995.

Janis, Irving and Leon Mann. *Decision-Making: A Psychological Analysis of Conflict, Choice, and Commitment*. New York: Free Press, 1979.

Jervis, Robert. *Perception and Misperception in International Politics*. Princeton: Princeton University Press, 1976.

Jervis, Robert. "Perceiving and Coping with Threat." In *Psychology and Deterrence*, edited by Robert Jervis, Richard Ned Lebow, and Janice Gross Stein. Baltimore: Johns Hopkins University Press, 1985.

Jervis, Robert. *System Effects: Complexity in Political and Social Life*. Princeton: Princeton University Press, 1997.

Jervis, Robert. *The Logic of Images in International Relations*. Princeton: Princeton University Press, 1970.

Kennedy, Paul, ed. *Grand Strategies in War and Peace*. New Haven: Yale University Press, 1991.

Kennedy, Paul. *Strategy and Diplomacy 1870–1945: Eight Studies*. London: Allen & Unwyn, 1983.

Kennedy, Paul. *The Rise and Fall of the Great Powers: Economic Change and Military Conflict from 1500 to 2000*. New York: Vintage Books, 1987.

Keynes, John Maynard. *The General Theory of Employment, Interest, and Money*. Basingstoke: Palgrave Macmillan, 2007.

Kowert, Paul and Jeffrey Legro. "Norms, Identity, and Their Limits: A Theoretical Reprise." In *The Culture of National Security*, edited by Peter Katzenstein. New York: Columbia University Press, 1996.

Larson, Deborah. *Origins of Containment: A Psychological Explanation*. Princeton: Princeton University Press, 1985.

Layne, Christopher. "From Preponderance to Offshore Balancing: America's Future Grand Strategy." *International Security* 22 (Summer 1997): 86–124.

Layton, Peter. "The Idea of Grand Strategy." *RUSI Journal* 157 (August 2012): 56–61.

Lebow, Richard Ned. *Between Peace and War: The Nature of International Crisis*. Baltimore: Johns Hopkins University Press, 1981.

LeDonne, John. *The Grand Strategy of the Russian Empire, 1650–1831*. Oxford: Oxford University Press, 2004.

Leffler, Melvyn. *A Preponderance of Power: National Security, the Truman Administration, and the Cold War*. Stanford: Stanford University Press, 1993.

Liddell Hart, Basil. *Strategy*. New York: Praeger, 1967.

Lippmann, Walter. *US Foreign Policy: Shield of the Republic*. Boston: Little, Brown and Company, 1943.

Luttwak, Edward. *Strategy: The Logic of War and Peace*. Cambridge: Harvard University Press, 2001.

Luttwak, Edward. *The Grand Strategy of the Byzantine Empire*. Cambridge: Harvard University Press, 2009.

Luttwak, Edward. *The Grand Strategy of the Roman Empire from the First Century A. D. to the Third*. Baltimore: Johns Hopkins University Press, 1976.

Luttwak, Edward. *The Grand Strategy of the Soviet Union*. London: Weidenfeld & Nicolson, 1983.

MacDonald, Paul and Joseph Parent. "Graceful Decline? The Surprising Success of Great Power Retrenchment." *International Security* 35 (Spring 2011): 7–44.

Martel, William. *Grand Strategy in Theory and Practice: The Need for an Effective American Foreign Policy*. New York: Cambridge University Press, 2015.

McDougall, Walter. "Can the United States Do Grand Strategy?" *Orbis* 54, no. 2 (2010): 165–84.

Mearsheimer, John. *Why Leaders Lie: The Truth About Lying in International Politics*. Oxford: Oxford University Press, 2013.

Meinecke, Friederich. *Machiavellism: The Idea of Raison d'État and Its Place in Modern History*. New Haven: Yale University Press, 1957.

Milevski, Lukas. *The Evolution of Grand Strategic Thought*. Oxford: Oxford University Press, 2016.

Mitzen, Jennifer. "Illusion or Intention? Talking Grand Strategy into Existence." *Security Studies* 24, no. 1 (2015): 61–94.

Murray, Williamson, Macgregor Knox, and Alvin Bernstein, eds. *The Making of Strategy: Rulers, States, and War*. Cambridge: Cambridge University Press, 1994.

Murray, Williamson, Richard Hart Sinnreich, and James Lacey, eds. *The Shaping of Grand Strategy: Policy, Diplomacy, and War*. Cambridge: Cambridge University Press, 2011.

Paine, SCM. *The Japanese Empire: Grand Strategy from the Meiji Restoration to the Pacific War*. New York: Cambridge University Press, 2017.

Parker, Geoffrey. *The Grand Strategy of Philip II*. New Haven: Yale University Press, 1998.

Popescu, Ionut. *Emergent Strategy and Grand Strategy: How American Presidents Succeed in Foreign Policy*. Baltimore: Johns Hopkins University Press, 2017.

Posen, Barry. "Command of the Commons: The Military Foundation of US Hegemony." *International Security* 28 (Summer 2003): 5–46.

Posen, Barry. *Restraint: A New Foundation for US Grand Strategy*. Ithaca: Cornell University Press, 2014.

Posen, Barry. *The Sources of Military Doctrine: France, Britain, and Germany between the World Wars*. Ithaca: Cornell University Press, 1984.

Putnam, Robert. "Diplomacy and Domestic Politics: The Logic of Two-Level Games." *International Organization* 42 (Summer 1988): 427–60.

Reich, Simon and Peter Dombrowski. *The End of Grand Strategy: US Maritime Operations in the 21st Century*. Ithaca: Cornell University Press, 2018.

Remnick, David. "Going the Distance: On and Off the Road with Barrack Obama." *The New Yorker*, January 27, 2014.

Ripsman, Norrin, Steven Lobell, and Jeffrey Taliaferro, eds. *Neoclassical Realism, The State, and Foreign Policy*. Cambridge: Cambridge University Press, 2009.

Ripsman, Norrin, Steven Lobell, and Jeffrey Taliaferro. *Neoclassical Realist Theory of International Politics*. New York: Oxford University Press, 2016.

Rosecrance, Richard and Arthur Stein, eds. *The Domestic Bases of Grand Strategy*. Ithaca: Cornell University Press, 2003.

Samuels, Richard. *Securing Japan: Tokyo's Grand Strategy and the Future of East Asia*. Ithaca: Cornell University Press, 2007.

Schweller, Randall. *Deadly Imbalances: Tripolarity and Hitler's Strategy of Conquest*. New York: Columbia University Press, 1998.

Schweller, Randall. "Managing the Rise of Great Powers: History and Theory." In *Engaging China: The Management of an Emerging Power*, edited by Alastair Iain Johnston and Robert Ross. New York: Routledge, 1999.

Silove, Nina. "Beyond the Buzzword: The Three Meanings of 'Grand Strategy.'" *Security Studies* 27, no. 1 (2018): 27–57.

Sprout, Harold and Margaret Sprout, "Environmental Factors in the Study of International Politics." *Journal of Conflict Resolution* 1, no. 4 (1957): 309–28.

Swaine, Michael and Ashley Tellis. *Interpreting China's Grand Strategy: Past, Present, and Future*. Santa Monica: RAND, 2000.

Tellis, Ashley, Janice Bially, Christopher Layne, and Mellissa McPherson. *Measuring National Power in the Postindustrial Age*. Santa Monica: Rand, 2003.

Trachtenberg, Marc. "Making Grand Strategy: The Early Cold War Experience in Retrospect." *SAIS Review* 19 (Winter 1999): 33–40.

Wolfers, Arnold. *Discord and Collaboration: Essays on International Politics*. Baltimore: Johns Hopkins Press, 1962.

Yetiv, Steve. *The Absence of Grand Strategy: The United States in the Persian Gulf, 1972–2005*. Baltimore: Johns Hopkins University Press, 2008.

Zakaria, Fareed. *From Wealth to Power: The Unusual Origins of America's World Role*. Princeton: Princeton University Press, 1998.

2   The great powers

For the purpose of this book, it is relevant to delve not only into grand strategy, but also into the topic of the great powers. What are the attributes a great power should have? Which states are presently great powers and why? Have great powers always existed in world politics or is the idea of great powerhood a recent invention? Who were the states that historically have been great powers? How do we know when a state becomes and when it ceases to be a great power? Finally, and most importantly, what are the criteria to classify great powers as either rising, status quo, or declining, and what are the general characteristics of each type?

The makings of a great power

Great power refers to a state that stands apart from others in three main ways, because of: a) power capabilities, b) behavior, and c) recognition by other actors.[1]

Great powers are states that own uncommonly large, sophisticated, and diversified capabilities. As discussed in Chapter One, these capabilities may be geopolitical, demographic, economic, military, technological, social, political, and diplomatic. A cursory look at the material indicators of power illustrated in Table 2.1, is telling.[2] One of the key reasons for the current componence of the great power club (the US, China, Russia, Great Britain, France, Germany, and Japan) is that all these states make the top ten of the world's largest economies as measured by GDP, as well as the top nine of the world's largest military spenders. Six of the current great powers—the exception being Russia, presently under economic sanctions for its annexation of Crimea—are among the 20 most innovative states in the world in terms of institutions, business, markets, infrastructure, and human capital. Five of the great powers make the list of the most populous 20 states in the world and the other two, Great Britain and France, are holding the ranks 21 and 22. Even in terms of territorial size, three of the largest four states in the world are great powers: Russia, the US, and China. All of the great powers minus France were in 2012 in the top ten states in the world measured by the most recent Composite Index of National Capability (CINC), a composite measure of the six indicators—military expenditures, military personnel, energy consumption, iron and steel production, total population, and urban population—employed by the Correlates of War project to assess the overall power level of states.[3]

Table 2.1 Material indicators of capabilities for the great powers

Great Power	Size of territory in square km (rank)	Population in millions (rank)	GDP in billions of dollars (rank)	Military spending in billions of dollars (rank)	Innovation index rank in the world	CINC (rank)
United States of America	9,833,517 (3)	329 (3)	21,482. 41 (1)	649 (1)	3	.139353 (2)
China	9,596,960 (4)	1,384 (1)	14,172. 20 (2)	250 (2)	13/14	.218117 (1)
Russia	17,098 (1)	142 (9)	1,649.21 (12)	61 (6)	46	.040079 (5)
Japan	377,915 (61)	126 (10)	5,220.57 (3)	46 (9)	15	.035588 (4)
Germany	357,022 (62)	80 (19)	4,117.07 (4)	49 (8)	9	.017911 (7)
France	643,801 (42)	67 (21)	2,844.70 (6)	63 (5)	16	.014207 (10)
Great Britain	243,610 (78)	65 (22)	2,809.91 (7)	50 (7)	5	.015277 (9)

Accordingly, great powers are not one-hit wonders, excelling only in one dimension while performing badly in the rest. As Kenneth Waltz writes, "states are not placed in the top rank because they excel in one way or another. Their rank depends on how they score on *all* of the following items: size of population and territory, resource endowment, economic capability, military strength, political stability, and competence."[4] This is why states that are doing well in isolated areas, but not in all, such as economic growth (Singapore, Qatar), technological innovation (Switzerland, Sweden), or nuclear weapons (Pakistan, North Korea), are not considered great powers. A great power must perform well across dimensions, even if its performance may be uneven. For instance, Great Britain from the 1830s to the 1890s deployed the world's largest navy, enjoyed the highest rate of economic productivity, and was by far the most developed state in terms of industry, but trailed behind other great powers in size of armed forces; meanwhile, Russia and the Soviet Union have traditionally fielded large armies, on shaky economic and technological foundations. However, all great powers stand apart from non-great powers not just based on the magnitude of their capabilities, but also on their diversity. Consequently, in order to rise to become great powers, states have to show improvement across the board, not only in a single area—militarily, economically, technologically, etc. By the same logic, great power decline is not triggered just by poor performance in one category, but reflects a deterioration of performance relative to the competition across dimensions.

It is also worth noting that the particular resources which matter most at a given time vary historically, depending both on objective technological and socio-economic trends, and on which dimensions states believe count as power at a given moment.[5] For instance, in pre-industrial societies, in which manpower and land were the primary factors in generating wealth and military force, territory and of population were considered critical. However, since the advent of the industrial age, the most important resources are economic productivity and military spending. Essentially, the state that is able to produce the most high-demand mass-produced goods and that can equip its military forces with the most advanced armaments is considered the most powerful. These assessments of importance of dimensions may very well shift again. Recent power comparisons between the US and China have downplayed the role of traditionally valued factors such as GDP, emphasizing, instead, alternative measures, such as inclusive wealth, which stress the state's infrastructure; technology; human capital including health and education, and natural capital including the environment, all of which generate wealth over the long run. In particular, technology, or the ability to research and develop cutting edge tools, machines, equipment, or techniques, may emerge as a future crucial dimension as the basis for obtaining a decisive advantage over rivals in hi-tech economic sectors and weapons.[6]

The power dimensions that are valued more, at any particular time, also depend on the intersubjective assessment of the states, reflecting their prevailing ideas, norms, and identities. For example, which of the military or economic capabilities are valued more will depend in part on the international and domestic views about the utility of war as an instrument of statecraft. If war is seen as a universal calamity not unlike an earthquake, from which no real winner can possibly emerge, states will value economic performance more. If, on the other hand, war is seen as a morally acceptable and cost-effective means to achieve greater security, prosperity, or national unity, the military dimension will be prized above the economic one. If a state, such as Japan or Germany after World War II, develops an identity as a pacific country, it will favor economic above military excellence, while a state that takes particular pride in the size and the past success of its army, such as Frederick II's Prussia, will value the military more.

Furthermore, whether great powers place greater value on naval or land military power may depend on their worldview. At the end of the nineteenth century, for example, most great powers swore by the writings of Captain Alfred T. Mahan, who made the case for the significance of naval power as the precondition for success in international politics. As a result, they developed large navies and began a hard-fought competition over the ownership of foreign naval bases and coaling stations. Similarly, in the 1950s and 1960s, the US and the Soviet Union engaged in a space race contest over launching satellites, sending a human being into space, and traveling to the Moon. This race was significant primarily because it was a symbolic reflection of supremacy in missiles, hence of the ability to deliver nuclear weapons, which was valued at the time ahead of the ability to field conventional forces.[7]

The rule of thumb in determining the importance of power dimensions, at any given time, is that great powers favor the economic and military dimensions above the rest, as wealth and force are not only the most easily measurable dimensions, but also the most direct instruments for advancing goals. This does not mean in any way that the other dimensions are unimportant, but rather that their importance stems primarily from their capacity to affect the effectiveness of exercising economic and military power. Thus, a large population or a strong capacity for technological innovation matters in as far as it is easily translatable into economic or/and military might.

To be a great power, a state must also act "like a great power should." The identity of being a great power comes loaded with expectations as to proper conduct: in other words, great powers play a distinct international role. This role follows the formula that "genuine great powers do x, but do not do y," which constrain any state claiming to be a great power to act/avoid acting in prescribed/proscribed ways.[8] Specifically, great powers are expected to control spheres of influence, assume responsibility for system management, maintain a high level of international activism, and either hold their own or prevail in confrontations with other states. Conversely, isolationism, passivity, shirking responsibility, lacking spheres of influence, and a string of defeats/setbacks at the hands of great powers and non-great powers are behaviors one does not expect to see from a great power.

A great power is expected to have a larger reach than the average state, in other words, to exert its influence over a wider geographical area and over a larger array of issues. Using Robert Dahl's categories, a great power surpasses other states' influence both in terms of domain (the number of actors it influences) and scope (the number of issues across which it exerts influence).[9] A typical state affects events solely in its immediate neighborhood, because the scarcity of its resources does not allow it to contemplate action both close to home, where such action is vital, and further afield, where it is not. By contrast, a great power has the resources to spare to affect actors not only in the region in which it is located, but also in several regions at once, or even the entire international system. The capacity and willingness to project power is the distinguishing criterion between regional powers that influence their own region only, for instance in present times Brazil, Turkey, Iran, Saudi Arabia, or South Africa; great powers that affect other regions as well as their own (France, Great Britain, Russia, China); and superpowers or world powers that affect the entire system (the US, the Soviet Union). In this sense, while the US may be a Western Hemisphere power located in North America, it is at the same time a European, East Asian, Middle Eastern, and Pacific power.[10]

Furthermore, an average state is able to pursue only a few basic objectives at a time, such as survival, territorial integrity, and maintenance of a modicum of prosperity, but will have few resources left for achieving more ambitious goals. Therefore, even if it wanted to, it could not deal with international issues that do not affect it substantially, but are of life and death to other members of the system, such as migration, drug trafficking, global warming, terrorism, or state aggression in another region. Nor could it afford the costs of promoting abroad its values, be they institutional, cultural, or religious; or of bolstering its international status. Great powers, however, can pursue such objectives, since their disproportionate share of capabilities shields them from the hand to mouth dilemmas common to other states.[11]

Great powers also assume a larger share of responsibility than the average state, meaning that they play a managerial role on behalf of the international system. Part of this task involves exercising international governance and part of it defending the international order. International relations cannot take place without some amount of stability and predictability, especially considering that there is no government as in domestic politics to enforce rights, such as ownership of territory and independence, or obligations, such as the duty to respect treaties. States would not be able to trust each other enough to conduct trade and investment, conclude alliances, or engage in diplomacy, unless they subscribe to certain shared rules of interaction. These rules however have to be managed, in the sense of having to be formulated, communicated, administered through the creation of caretaker institutions, interpreted, enforced, and legitimized in the eyes of states. Furthermore, for the rules to work at all, there have to be actors that intervene in order to prevent and limit conflict among states, and in particular, crises and war from getting out of control to the point of spreading to the entire system, through diplomacy, mediation, sanctions, and, if need be, use of force.[12] As seen in the above, the only states that have the necessary resources and the will to take on these functions are the great powers. To be sure, great powers extract commensurable benefits from their playing the part of rule-makers and defenders of the international order, but also have to foot the management bill on behalf of the other states.

Carrying out systemic managerial duties requires that any would-be great power demonstrate a continual propensity for using force and extending commitments to come to the defense of other states. A telltale sign that a state is a great power consists in a high level of participation in wars and alliances. There is strong statistical evidence that in the last 200 years, war participation and great power status went hand in hand. In this timeframe, 95 interstate wars were fought, of which 55 involved at least one great power, which makes great powers account for nearly 58% of the total. The figure is even greater for wars great powers fought against non-state actors, whether colonies or entities failing to meet the criteria of statehood, i.e., non-European polities in the nineteenth century such as the Zulu Empire, or in the present, the Taliban, the Islamic State, or Al Qaeda: 67% or 110 out of 163 extra-state wars involved great powers. Meanwhile, from 1815 to 2008, there have been 204 alliances involving commitments to defend the other side, of which 181 involved at least one great power, representing 88% of the total. Scholarship constantly shows that once a state is recognized as a great power, its war involvement goes up, while a corresponding drop in war participation accompanies its exit from great power ranks—think about the increase in US war participation since 1898 when the US became a great power, as opposed to the drop in Sweden's from 1721, when it exited the great power club.[13]

Lastly, a key behavioral trait for great powers consists of proving successful in wars, crises, and diplomatic confrontations, both by being able to win when pitted against non-great powers, and by putting up a strong fight against other members of the great power club. In effect, the traditional test for a state becoming a great power has been success displayed in war, crises, and diplomatic disputes. The origins of this view are to be found with the nineteenth-century German historian Leopold von Ranke, who popularized the study of history as the study of the great powers. According to Ranke, "one could establish a definition of a great power that it must be able to maintain itself against all others, even when united." However, this definition exaggerates, since usually most great powers fight either isolated enemies or are themselves part of coalitions. This is why this requirement is more commonly interpreted as demanding that great powers hold their own when pitted against another great power. As Jack Levy puts it, "a Great Power is one that can afford to take on any other power whatever in single combat."[14] By implication, great powers are also expected to prevail when confronting lesser opponents that should not be able in turn to hold their own against them in one-on-one combat. To quote A. J. P. Taylor, "great powers were, as the name implies organizations for power, that is, in the last resort, for war. They might have other objects. ... But the basic test for them as Great Powers was their ability to wage war." As will be shown below, the point of induction of new members in the great power club has coincided in almost every single instance since 1648 with its successful prosecution of a war against another great power. Similarly, as Martin Wight writes, great powers rarely die of old age in their beds. Most often their exit from the great power club comes as a result of a crushing defeat/occupation at the hands of other great powers from which they are unable to recover.[15]

As attested by Russia's Foreign Minister Alexander Gorchakov's (1856–1882) famous quip, great powers are not recognized but reveal themselves, by which he meant through war. But, while success is critical for recognition as a great power, it is not the only criterion that matters for great powerhood. Consequently, defeating a great power without exhibiting the other traits required of a great power does not make the victor a great power, as seen in the cases of Vietnam or Afghanistan, which could claim victory over the superpowers. By the same token, a loss to a non-great power does not mean automatic demotion. As Hans Morgenthau suggested, great powers are like large banks, with enough resources to spare to afford losses: "a bank with large, proven resources and a record of successes can afford ... to make a mistake or suffer a setback."[16] Hence, if the great power conserves sizable resources, as was the case with the US after Vietnam, one defeat is not enough to exclude it from the great powers' ranks. However, great powers cannot go on an uninterrupted string of defeats without raising questions as to their proper credentials. To quote President Kennedy, "there are limits to the amount of defeats I can accept within a twelve-month period. I've had the Bay of Pigs, and pulling out of Laos, I cannot accept a third."[17] To put an end to these successive setbacks, it was important for Kennedy to make a stand over South Vietnam.

Assuming therefore that success matters a great deal, how does one evaluate success in world politics? Several points are worth considering. First, success refers to goal attainment or how close a great power manages to translate its objectives into practice. To achieve victory against another state means, as Clausewitz suggested, to impose one's political will against the resistance of the adversary. Second, if success is not viewed just as goal attainment, but also as score-keeping, great powers should be wary of the danger of Pyrrhic victory. In other words, achieving one's objective, but at a ruinous cost, offsets benefits or makes the great power vulnerable in a repeated or fresh future contest. Therefore, for a state to be successful, it has to be able to obtain victory in a cost-effective manner. Gains have not only to exceed, but also to make up for the costs. Third, success can be evaluated against existing benchmarks, which do not only have to do with stated goals, but also with the expectations of a state's likely performance. If, for instance, a perceived strong actor is expected to win easily (for instance China against Vietnam in 1979, or the US against insurgents in Iraq from 2003 to 2008), but can only achieve victory after a hard-fought and costly struggle, its success is questionable, as it falls behind the initial expectations. Conversely, if a weak actor performs better than expected, for instance fighting a great power to a standstill, it achieves success just by managing to avoid loss (the Mujahedeen against the Soviet Union). Fourth, success is not always clear-cut, and, as such, it is susceptible to manipulation. This is particularly the case of crises and disputes that do not escalate to war and where the outcome remains ambiguous since there is no defeated party for everyone to see. Therefore, considering the advantage conferred by success, it is not uncommon for both sides to claim victory at once, which is an effort to manipulate both domestic and international public opinion into accepting their claim, thus "fixing" the result. That is to say, success can also be a matter of interpretation and perception in non-decisive contests. To the extent that an actor is believed to have won, it has de facto managed to win.[18] Success will be attributed in such cases to the state that manages to present the most persuasive argument to a broad international audience based on how it has met the above-mentioned criteria (goal attainment, score-keeping, and expectations) better than the other side.

The third and final condition of being a great power is recognition by other actors. Recognition may seem problematic at first glance, since it rests on an ostensible tautology: a state is a great power to the extent that other great powers see it as a great power. Yet, this circular requirement may be understood through a metaphor: imagine an elitist social club, where the members vote on whether to admit a new applicant. It is in this sense that a state becomes a great power only after the established club members accept it as one of them. Of course, the club has not always been in existence—and for its original founders, states such as France, Austria, England, and Spain, there was no such admission requirement. But every new member that has sought to join since had to submit itself to the vetting of the existing membership.

Recognition represents the other side of the coin of the requirement of great power behavior. It is not enough for a state to behave as a great power ought to. The other states as well have to behave toward it as if indeed it were a great power. Social conventions regulate how great powers should be treated by their presumptive peers, as well as by ordinary states. Thus, Levy writes that being a great power supposes being formally included in major international conferences, congresses, treaties, and international institutions, as well as being treated as an equal regarding rights, respect, and protocol by the other great powers.[19] Basically, this means that special consideration should be shown to the interests of fellow great powers; and that equal or superior deference should be shown toward their status.

Great powers should avoid treading on each other's interests. The reasons are that challenging a great power's interests risks provoking conflict with the affected party and, moreover, that other great powers may intervene as well in the resulting conflict, causing an expensive and lengthy system-wide confrontation. Hence, to avoid these outcomes, great powers subscribe to a common code of conduct akin to a gentleman's agreement, whose main rule is that a great power should not reach important international decisions without previously consulting the other members of the club. Consultation may be either formal, in the setting of an international social gathering such as a congress, conference, or the regular meetings of an international organization; or informal, taking place through bilateral diplomatic contacts. The point of consultation is allowing the power that is consulted to express objections or counterclaims to another's planned action before it occurs. Therefore, the consulted power is able to warn the other great power not to proceed, or to prompt it to offer concessions in return for its support or neutrality.

Accordingly, great powers have devised throughout history a multitude of mechanisms for avoiding unnecessary conflict with other great powers. For instance, if a great power gained a strategic advantage over its rivals, especially in terms of territorial possessions, the other great powers were entitled to compensation by securing a "proportional aggrandizement" in other quarters. Conversely, if a great power suffered a loss, it could expect an indemnity from the other club members to make up for its disadvantage. Another solution was to draw explicit spheres of influence accords, by which a great power marked the boundaries of its geographical interests, excluding other great powers from the respective area in return for a similar acceptance of the boundaries established by them. Finally, great powers found it expedient to neutralize areas of contention that separated their spheres of influence, hence, the term "buffer state" used to designate such intermediary areas. Several contiguous buffer states that prevented spheres of influence from touching upon each other constituted a so-called cordon sanitaire.[20]

This special consideration extended to the interests of other great powers remains a valid rule up to today. Actually, the key function of the institution of the veto for the permanent members of the UN Security Council (the US, Great Britain, Russia, France, and China) is to allow each of them to withdraw from the Council's debate any decisions that could hurt its interests.

Similar solicitude is not extended by great powers to non-great powers, because even if they were to show hostility toward a great power's plans, they would be unable, due to their relative weakness, to trigger a major conflict. To this extent, being regularly excluded or left out from consultations is an indication that the state is not important enough to be taken into account (or a *quantité négligeable*— a negligible amount), and, as such, is not considered a great power by the club members.[21] By contrast, being included conveys the opposite message: that the state cannot be neglected and that it has become a recurrent factor of concern in the strategic calculations of the great powers, which therefore acknowledge it to be just as significant as they are.[22]

Great powers, however, do not care just about their material interests. They also care about their status, which is to say about the respect due to them. This is not silly vanity. As Levy writes, "symbolic interests of national honor and prestige are ... given high priority by the Great Powers, for these are perceived as being essential components of national power and necessary for Great Power Status."[23] Status refers to social rank in the hierarchy of states, which makes the great powers the top actors. Status is important for two reasons. First, it is essential for achieving and maintaining material perks and political influence. This is so because status imposes compulsory deference on the part of the lower ranks toward the upper ranks in the form of granting them various privileges, priorities, immunities, attention, and exemptions from common obligations. Such deferential behavior is often symbolic. For instance, someone enjoys superior status if he or she is allowed to enter, exit, pass, sit, or speak first, or is addressed with particular honorific terms in speaking or writing ("Mr. President," "your honor," "your majesty"), is given more attention when speaking (there are no admissible interruptions, and contradictions have to be softened), or is permitted exclusive access to a certain status symbol such as special food, clothing, housing, transportation, or education (for instance designer clothes, traveling by limousine or private plane, attending an Ivy League school). While great powers may obtain corresponding rewards from other states by resorting to threats or bribery, it is more cost-efficient to rely on status, because it is freely granted as an entitlement due to social prominence. Second, status also matters as a means for achieving and maintaining psychological self-esteem by being publicly recognized as better and worthier than other actors. This is true for both individuals and states. Research has conclusively proven that individuals care about how their group is appreciated relative to other groups, even when there is no contact with other members of the group, or the group has been put together randomly. Moreover, states are led by individuals who are not only members of the group, but are also visible symbols of the group. Even if elites do not care about status themselves, they are compelled by their respective domestic audiences to pay close attention to the standing of the state vis-à-vis the other states in the system.[24]

The point is that great powers require being treated at all times in a particular deferential manner by other states, peers, and non-peers. Withdrawing or failing to extend such deference represents an insult. The meaning of an insult is to humiliate symbolically the offended party, or to assert its inferiority by relation to the offender. This can be done by breaking protocol, or by provoking or daring the insulted to a public test whose goal is to check whether it really possesses the qualities that are the basis of its claim to high status. Even if such behaviors are symbolic, their meaning is clear—to call into question the right to high status of the offended. They are the equivalent of asking it "who are you that I should obey/pay attention to you?" Such insults require a response on the part of the injured party, because the lack of reaction or the refusal to take part in the proposed challenge after being dared represents a de facto confirmation of the charge leveled, i.e., that the insulted party does not in fact deserve high status.[25]

Two consequences follow from here. First, in order to avoid conflict, other actors, whether great powers or non-great powers, need to exercise considerable caution in extending at all times proper deferential treatment. This is not to say that great powers will go on the warpath at every minor faux-pas in protocol, but other actors must pay attention to avoid placing them in an untenable situation where status concerns will force them either to fight/keep fighting or to suffer a major public loss of face. As such, the rule of thumb in dealing with a great power is to avoid humiliating it in any circumstances, which would necessarily provoke its hostility.[26] It is for this reason that even if a great power were to suffer a defeat or setback, the victor should continue to extend it the same respect due to any other social peer or superior. As long as a state keeps commanding such respect, it remains a member of the great power club.

Second, a state claiming to be a great power cannot afford not to care whether it is shown proper respect by other states. A great power considers its status or reputation an essential national interest, on par with its independence or territorial integrity. Therefore, if a state has made a claim for deferential treatment, and then fails to enforce it by refusing to fight for it, this shows it is not a fully-fledged great power. As the German historian Heinrich von Treitschke wrote:

> whoever attacks the honour of a state … thereby impugns the essential character of the state … [a state] cannot permit this power to be questioned even symbolically … if the flag of the state is insulted, it is the duty of the state to demand satisfaction, and if the satisfaction is not forthcoming to declare war, however trivial the occasion may appear for the state must strain every nerve to preserve for itself the respect it enjoys in the international system.

This position is by no means anachronous, as evidenced by Henry Kissinger, who expressed the opinion that "no serious policymaker could allow himself to succumb to the fashionable debunking of 'prestige,' or 'honor,' or 'credibility.'"[27] In fact, few serious policymakers do. A recent lab experiment involving 77 actual military and political leaders confirmed that group status loss represented a major factor in their decisions. In total, a study of wars from 1648 to the present found that status represented a principal cause in more than 58% or 62 of the 94 wars fought since 1648, constituting "by far the leading motive" to resort to force. [28]

All the above conditions—capabilities, behavior, and recognition—are necessary and none on its own is sufficient for a state to be identified as a great power. A state may have great capabilities and yet be denied recognition as a great power. This is the case of rising powers in the early stages of their rise. These states have resources that are essentially on par with those of the great powers, but may fail to act as expected of a great power, for instance by developing spheres of influence, or by assuming managerial duties on behalf of the international system (for instance, the US from the 1860s to 1890s). Or the fledgling rising power may fail to be acknowledged as an authentic member of the great power club by the existing members and by the other states in the system. This may be due partly to the gap between the actual shift in capabilities and the perception by other states that such a shift has occurred. As Lord Bolingbroke remarked,

> the precise point at which the scales of power turn is imperceptible to common observation. ... They who are in the rising scale do not immediately feel their strength, nor assume that confidence in it which successful experience gives them afterward. They who are the most concerned to watch the variations of this balance misjudge often in the same manner. ... They continue to have no apprehensions of a power that grows daily more formidable.[29]

However, this may also be due to the fact that club members withhold recognition for so long as the rising power has not proven its credentials by successfully standing up/defeating an acknowledged great power (for instance, Japan in the late 1890s after it defeated China).[30] Similarly, some of the strong regional powers may on occasion behave in ways characteristic of great powers (for instance, present-day Turkey and Brazil in brokering talks on lifting sanctions on Iran, a managerial duty; or Turkey intervening out of region in Libya), or, as noted above, might even score victories over the great powers. However, in the absence of capabilities on par with the existing powers and without being treated as a great power should, they will not be accepted as great powers.

It is rarer for a state to be recognized as a great power, without being at the same time a great power. Yet, as will be shown below in the cases of the Ottoman Empire and Imperial China, former powerhouses that had been dominant in their own regional systems but which were never properly integrated in the great power club, it is possible for a state to be treated on occasion by others as if it were a great power, because doing so confers certain temporary political advantages. However, unless the state possesses genuine great power capabilities and behaves in line with great power standards, it will not be accepted as one. Thus, being treated as a quasi-great power does not translate into lasting privileges like those enjoyed by club members in good standing, nor makes the state a regular factor affecting the other great powers' grand strategy calculations.[31]

The origins of the great powers

When should an investigation into the grand strategies of great powers begin? Although every historical era and international system have witnessed the presence of states stronger than the rest in capabilities, such as Pericles' Athens, Imperial China, the Roman Empire, or the Habsburg Empire of Charles V, it may be anachronistic to consider them great powers in the sense employed above. These actors may have fulfilled the first criterion—power capabilities—but they had little notion of the behavior expected from a great power and of the recognition a great power expects from other states. Accordingly, great powers in the modern sense have emerged in world politics only relatively recently.

But when exactly was that? Levy argues that great powers emerged at the end of the fifteenth century. A host of practices and ideas related to great power politics first came into scene around 1494, the year France invaded Renaissance Italy, among them, diplomacy, sovereignty, the balance of power, raison d'état, rivalry, as well as organized gunpowder warfare. However, the concept of great power only came into explicit use in 1814, when during the Congress of Vienna that followed the Napoleonic Wars, there was a formal move to restrict decisions to the five "Powers of the first order," who each could put in the field at least 60,000 troops.[32] This book settles however for a different date for the emergence of great powers: the Treaties of Westphalia of 1648.

Traditionally, the two Treaties, signed at Münster and Osnabrück, have been regarded as the moment of birth of the modern sovereign state system, based on the widespread interpretation that they allowed the German states that were part of the Holy Roman Empire, a loose federation not unlike the present-day EU, to conduct independent foreign relations. Recently, modern scholarship has exposed this traditional view as a myth, pointing out that the German states had enjoyed the right to conclude alliances outside the Empire since well before the Peace of Westphalia, and that, even in the aftermath of the Treaties, there was still little notion among them of genuine domestic or external sovereignty.[33] Nonetheless, despite these findings, Westphalia is still the most likely date for the birth of great powers for several reasons. First, the Treaties marked the end of the Thirty Years War, which was a general conflict that involved, at some time or another, every major state in Europe (England, that had undergone a civil war, was the only one not to fight, but had an important part to play in the war's settlement). This pattern of system-wide warfare has become since then a recurrent feature in international politics, taking place periodically. To mention only the most important of such conflicts, the end of the seventeenth century and start of the eighteenth century had the wars of Louis XIV; the mid-eighteenth century had the Seven Years' War; the cusp of the nineteenth century had the wars of the French Revolution and of Napoleon; and the twentieth century had the two world wars. But, as Levy admits, there is no previous instance of this pattern prior to the Thirty Years' War—not in the Hundred Years' War, or even in the wars of Charles V or Philip II, which pitted the Habsburg Empire against France, England, or the Ottoman Empire, but did not provoke general warfare. Great powers thus fought localized duels against each other, without involving states further afield.[34]

Second, the Congress of Westphalia, a diplomatic meeting that ran over seven years and eventually produced the Treaties, was also the first event of its kind. With 176 plenipotentiaries representing 194 European rulers, it was the largest peace conference that had ever been assembled up to that point. It inaugurated the practice of universal-scale conferences, where nearly all states, and more importantly all key players, formally met to decide principles and practical features of international order in the aftermath of a general war. This practice has been followed ever since, with successive such congresses, notably the ones taking place in Utrecht in 1713, Vienna in 1814–1815, Versailles in 1918, and San Francisco in 1945.[35] But there is no comparable diplomatic reunion prior to Westphalia.

Third, building on the first two points, system-wide participation in both matters of war and of peace shows beyond any doubt that these matters affected all great powers, constituting an unmistakable admission of the fact that they were inextricably linked to each other. In other words, Westphalia marked the transition from a number of separate regional systems whose conflicts and resolutions of conflicts were pursued independently, primarily because they were thought to bear no connection to each other, to an integrated international system. The very definition of an international system is "the ensemble constituted by political units that maintain regular relations with each other and that are all capable of being implicated in a generalized war." Therefore, in order for an international system of the sort to exist, it is necessary for the actions of one party to have reverberations for the whole, so that "the behavior of each is a necessary factor in the calculations of others."[36] Accordingly, great powers as strategic actors, calculating the impact of their actions on others, the response of the other side, and the way that actions of third parties could affect them, could only exist and function in the sort of international system that first emerged in 1648.[37]

Neither the end of the fifteenth century nor the end of the nineteenth century are convincing dates for the origins of the great powers. True, the appearance of modern diplomacy can be traced to the emergence of the office of the ambassador as a permanent representative of the state in Renaissance Italy, which is why most diplomatic histories begin their accounts around that time.[38] Even so, the conduct of diplomacy did not suppose for centuries any bureaucratic supervision or organization even in the case of the strongest diplomatic powers. When the Cardinal de Richelieu became the Prime Minister of France in 1624, he had to beg the French envoys to send him copies of their letters of instruction, since no such records had been previously kept in Paris; and the entire French foreign office at the time of accession of Louis the XIV as king consisted of one secretary and five clerks. Thus, the ideas and practices mentioned by Levy were only in their infancy at the time, in the sense that they were limited both in the extent and the length of time to which they were pursued. Therefore, they became the prevalent norm only much later, typically in the interval from the Treaties of Westphalia to the wars of Louis XIV, in other words, in the second half of the seventeenth century. In effect, sovereignty, as seen in the above, in the sense of a supreme authority in domestic and foreign affairs, had not been firmly accepted even in the aftermath of Westphalia 150 years later—it became a mainstay principle only around the time of the Utrecht Treaty of 1713. Concerning balance of power, although the Italian states concluded a coalition to balance an invading France in 1495, their arrangement lasted only four years, after which the fulcrum of the coalition, Venice, changed camps and joined France so as to despoil its former ally Milan. States might have been occasionally spurred by power rivalries, but these considerations often continued to be superseded by religious/ideological considerations. Spain and France, for instance, pursued a decades-long struggle over the control of Italy, which eventually ended in 1559 with France completely rallying behind Spain, a fellow Catholic power whose help was needed in order to repress Protestantism at home. Armies did employ gunpowder weapons such as the arquebus and the musket, but firearms were only used in conjunction with traditionally armed troops relying on weapons such as the pike. The proportion of shot to pike in the army was, through the sixteenth century, about 1:3, and it only reached a ratio of 3–4:1 around 1650. It was during the Thirty Years' War at the Battle of Breitenfeld (1631), which pitted Sweden against the Imperial army, that the superiority of troops armed with firearms was conclusively proven—the musket consequently becoming "the queen of the battlefield" for the next two centuries. Accordingly, military historians argue that the international system underwent a military revolution throughout the sixteenth and seventeenth centuries, reaching its apex in the latter as evidenced by changes in military tactics and strategy, increase of army size from tens to hundreds of thousands of men, and the multiplication of the costs of mobilizing society for war.[39]

Conversely, if 1494 appears as too early for the birth of great powers, 1815 appears as too late. For more than a century and a half, and especially throughout the eighteenth century, the classic age of power politics, the statesmen and scholars in Europe had acknowledged not only de facto but in explicit documents the existence of a class of states above the average. These states were considered to follow permanent interests which were calculated by taking into account the power and status of other states, as well as by anticipating the likely reaction of these other parties to their initiatives.[40] Thus, when the concept of great power was first put forward in 1815, it reflected a reality already more a century and a half in the making.

Identifying the great powers

How do we know when a state has become a great power and, conversely, when it has stopped being a great power? These questions are important not only for identifying which states are great powers, but also for determining their length of tenure, that is to say their point of entry and exit from the great power club.

This book argues that the entry and the exit into the great power club coincide with the extension and withdrawal of recognition. Hence, a state becomes a great power once it is formally accepted as such by the existing club members, provided that the criteria of capabilities and behavior are also met. As seen in the above, this usually happens once the state in question has proven itself in a major confrontation against an established great power. Russia became a great power by defeating Sweden in the Great Northern War (1700–1721). Prussia demonstrated its great power credentials by defeating Austria in the War of Austrian Succession (1740–1745) and standing up to Austria, France, and Russia in the Seven Years' War (1756–1763). Italy, formerly the kingdom of Sardinia, was acknowledged as a great power when it achieved unification after pursuing in conjunction with France a successful war against Austria (1859–1860). Japan was recognized as a great power after its defeat of Russia, in the Russo-Japanese War (1904–1905). China was already extended the status of a great power complete with a permanent position in the UN Security Council for its war participation against Japan in World War II, but it only became a genuine great power once it unified in 1949 and then went on to fight the US to a standstill in the Korean War (1950–1953). Only the US is a partial exception to this rule—defeating a former great power (but arguably quasi-power) Spain in the Spanish-American War (1898), which, nevertheless, still counts as an instance of a great power gaining recognition through victory on the battlefield.

By the same logic, a great power ceases being a great power once it loses recognition from other states in the system as a consequence of a lasting defeat to another great power. Sweden exited the club following its defeat and dismemberment in 1721. Slightly more controversially, Holland's inability to put up a fight to defend the so-called Dutch barrier of fortresses protecting the Austrian Netherlands (present-day Belgium) from France in the War of the Austrian Succession marked its demotion from great power ranks. Spain never recovered from its defeat to Great Britain in the Seven Years' War and was eventually occupied by France in 1808. Austria's nearly three centuries' tenure as a great power ended with the defeat and breakdown of its empire over much of Central Europe in 1918. Finally, Italy's ambitions as a great power ended with its defeat and occupation by allied forces during World War II in 1943.

Of course, it could be pointed out that many more great powers than the ones mentioned in the above have suffered catastrophic defeats or experienced periods of severe internal weakness—for instance, France in 1815, 1870, and 1940; Germany and Japan in World War II; or Russia after the end of the Cold War. Did they cease to be great powers?

Two points of view exist on the matter. According to the first viewpoint, represented by the Correlates of War dataset, great powers regularly drop out of the club in the aftermath of particularly harsh defeats, internal turmoil, or/and possible occupation by another state or states. However, the demoted power often reasserts its status at a later date. Consider for instance France. According to the Correlates of War, France lost its great power status on three occasions, as a result of defeats and occupation by enemy forces in the Napoleonic Wars (1814–1815), the Franco-Prussian War (1870–1871), and World War II, only to regain it each time. However, this book prefers the second viewpoint, argued by Levy, according to whom a state continues to remain a great power even if it is temporarily weakened, defeated, or occupied, provided that it manages to recover within "a reasonable period of time." Preference is given to this view because interruptions in great power tenure due to defeat, weakness, or occupation are too frequent to be of lasting significance, particularly since in most cases great powers conserve sufficient latent resources to make a speedy comeback.[41] While Levy does not specify what sort of interval would constitute a reasonable period, this book proceeds from the assumption that this interval should be set to 25 years, essentially a quarter century after suffering defeat/ turmoil/occupation. Hence, based on this interpretation, a great power does not cease immediately being viewed as a great power following a major setback, but only if it does not manage to make a recovery in the allotted period of time. If recovery fails to occur, the great power's exit is dated from the time of its defeat. Consequently, to again refer to the example of France, instead of several separate stints as a great power, the state is considered a great power uninterruptedly from 1648 to the present, based on its successful recoveries after each major setback. A similar logic is applied to the cases of Germany and Japan, which recovered within a quarter century of their defeat and occupation in World War II; and of the Soviet Union and Russia, which made a comeback after defeat in World War I (1914–1918) and internal turmoil following the end of the Cold War. By contrast, the defeats of Sweden, Spain, Holland, Austria, and Italy were definitive: there was no recovery in their case.

Accordingly, based on these criteria since the days of Westphalia there have been 12 states to have been acknowledged as great powers, as illustrated by Table 2.2.

The founders of the great power club were the major states that fought the Thirty Years' War—France, Sweden, and Holland on the side of the winners, Spain and Austria on the losing side—plus Great Britain. Only six great powers have joined the great power club since: Russia and Prussia in the eighteenth century, Italy and the US in the nineteenth century, and Japan and China in the twentieth. Five great powers dropped out of the club: Sweden, Holland, and Spain in the eighteenth century, and Austria and Italy in the twentieth. Consequently, seven great powers presently compose the great power club—France and Great Britain have been members since its founding, and, in order of their joining the club, Russia, Germany, the US, Japan, and China.[42] The composition of this list is constant among scholars, with occasional additions (the Ottoman Empire, Poland, and Denmark for the seventeenth century, or the EU and India for the present), but with few of the above states missing from any of the major accounts.[43]

The two omissions from the great power list that may require justification given the timeframe of the present investigation involve the Ottoman Empire and Imperial (Qing) China. Both states possessed large demographic resources and geopolitical holdings, which could translate into significant economic productivity and military forces. Actually, in terms of manufacturing output, China shows up ahead of every other great power except Britain and the US

Table 2.2 Composition of the great power club, 1648–2018

No.	Great Power	Entry	Exit	Great Power Tenure Length
1.	France	1648		370 years
2.	England/Great Britain	1648		370 years
3.	Austria/Austria-Hungary	1648	1918	270 years
4.	Spain	1648	1763	115 years
5.	Sweden	1648	1721	73 years
6.	Holland	1648	1748	100 years
7.	Russia/Soviet Union	1721		297 years
8.	Prussia/Germany	1745		273 years
9.	Italy	1860	1943	83 years
10.	United States of America	1898		120 years
11.	Japan	1905		113 years
12.	China	1950		68 years

even up to the 1880s.[44] Accordingly, one may follow one of two possible ways to proceed.

The first is to admit that both the Ottoman Empire and Imperial China counted as great powers, albeit in later centuries as declining ones, due to their extensive capabilities, spheres of influence, interests, and activism in the sense of frequency of participation in wars. Both were eventually excluded from the club because of structural weakness and successive defeats at the hands of the European powers and Japan. This view is usually expressed concerning the Ottoman Empire, which is seen as a great power until its defeat in the Austro-Ottoman War of 1683–1699 and the resulting peace treaty of Carlowitz, which forced it to give up for good Hungary and Transylvania. If so, the same principle should also apply to Imperial China, up to the Second Opium War (1856–1860), or even the Sino-Japanese War (1894–1895).[45] The second course of action is to consider the Ottoman Empire and Imperial China (maybe together with Spain in the nineteenth century) as quasi-powers, states which for limited intervals and for specific reasons were extended treatment commensurate with that of great powers, but that never satisfied the full admission criteria or were extended great power treatment on a permanent basis. This means that this treatment was unrelated to the states' capabilities or behavior; and was strictly determined by political expediency, for instance Germany treating the Ottoman Empire as a great power to secure its alliance prior to World War I, or the US relegating China as one of the four world policemen during World War II.

The motives as to why the latter interpretation may be preferable are twofold. First, the Ottoman Empire and China were never accepted as fully-fledged members of the state system dominated by the states of Europe even at the height of their capabilities. Ostensibly, the roots of this rejection had to do with deep-seated religious, cultural, and racial differences. Hence, not being part of the system, the Ottoman Empire and Imperial China could not have been admitted among the members of the great power club, regardless of the state of their capabilities. Even at the height of their power, neither state was extended on a permanent basis a treatment corresponding to a fellow great power by being consulted in major decisions, or by being allowed special privileges and respect.[46]

But another important motive behind the Western powers' refusal to acknowledge the two states as great powers was that they were not interested in securing such recognition. In other words, both the Ottoman Empire and Imperial China refused for a considerable length of time to play the great power game by the Western rules. Instead, they each considered themselves, due to their resources and accomplishments, as the natural world leader, and not one of many great powers. As such, they rejected a framework in which each great power manages the international system in conjunction with the other members, and, as a result, resorts to calculations of its grand strategy in accordance with the likely moves of the other players. While the Western powers were sending foreign envoys to Beijing and Istanbul, even stationing permanent embassies in the latter, neither Imperial China nor the Ottoman Empire sought to cultivate reciprocal diplomatic relations with them. Thus, China refused to engage in any foreign relations that did not acknowledge symbolically its supremacy through the offering of tribute or the performing of the kowtow to the emperor, a ritual of nine bows and nine prostrations. Meanwhile, the Ottoman Empire did not consider necessary until the 1790s to send permanent embassies abroad.[47]

Rising, status quo, and declining powers

Not all great powers are alike. Although great powers stand apart from the average state in the international system, they do not behave in the same way, and, as such, they differ considerably in their grand strategies. There are three basic grand strategy great power archetypes, pertaining to rising powers, status quo powers, and declining powers.[48] What determines whether a great power is rising, status quo, or declining, and how can they be told apart? In a nutshell, a great power's type is determined by the evolving trajectory of both its power resources and status rank. Rising powers are states whose levels of capabilities improve relative to those of acknowledged great powers. They seek to ascend to a higher rank than they presently occupy in the international pecking order. Status quo powers are powers whose levels of resources remain roughly on par or moderately improve by comparison with other great powers. They seek to maintain or marginally better their existing status. Finally, declining powers are powers whose levels of resources diminish relative to other great powers. They seek to avoid demotion to a lower rung than they presently occupy, or having already suffered demotion, seek to regain their former status. It is important to note that the evolution of a great power's capabilities and status should occur over time, in the sense that fluctuations that are quickly reversed do not count as ascending, descending, or maintaining-momentum long-term trends. Furthermore, what matters is not just objective trends of capabilities going up or down, but also, since status is involved, the perceptions and self-perceptions of where a great power ranks at a given time.

Rising power designates an already well-endowed state, whose perceived progress in terms of power is on the brink of elevating to the next rung in the great power hierarchy. Attaining each superior rung requires that the rising power should be perceived as possessing a varied portfolio of capabilities, and demonstrating great power behavior, including a proven record of military achievements versus the existing great powers. Progress in isolated areas is not sufficient for a state to be a rising power—the state needs to do well in all power dimensions. Moreover, to rise is not necessarily to overtake, but rather to narrow the gap with the members of the club to the point where performance becomes roughly comparable. A state may considerably improve its position relative to the other great powers or the number one, even surpassing the latter in single dimensions, without necessarily displacing its competitors from their overall superior position, as was the case for some of the rising great powers (Italy, US, Japan) in the nineteenth century. If a rising power's claim for advancement is recognized, its rise has been completed by becoming a club member, a top contender, or a new number one.[49]

Therefore, rising power is a label that fits several kinds of states. One kind of such great powers ("newcomers") consists of strong regional states who seek to gain admission to the great power club. Accordingly, every new state that has joined the great power club since 1648 has started up as a rising power. Another kind of rising powers ("aspirants") consists of states that have made it into the great power club, but who experience further growth, and seek to climb from the bottom ranks to achieve the position of frontrunner or first-rate power (ranks two to three in the club). This might be the case of contemporary China, which rose from the condition of arguably the weakest member of the club in the 1970s to the second economic and military power in the world in the present decade. The third type of rising powers ("successors") consists of states that, having achieved a frontrunner rank, continue to grow and consequently seek to supplant the existing number one as the dominant state in the system. Examples may be Imperial Germany in the early 1900s, and the Soviet Union during the Cold War.

This typology would result in the following list of rising powers from 1648 to the present day, as illustrated in Table 2.3.[50]

Currently strong regional powers, such as India, Brazil, South Africa, and Turkey, that are sometimes referred to as rising powers, do not make the list yet for several reasons. First, with the exception of India, none has yet sufficiently diverse capabilities that could rival those of states in the great power club. Second, even in the case of India and Turkey, none of them behaves as great powers do. They lack extra-regional interests or a sphere of influence; they are unable or refrain from playing a part in the management of the system; and they have not demonstrated their ability to hold their own in a conflict against a member of the great power club. Third, none are recognized as great powers either by being a regular part of the great powers' strategic calculations, or by being extended deference appropriate to a great power. Out of the four candidates, India appears the best placed to become eventually a rising power.[51]

Status quo power points to a state which is already a bona fide member of the great power club and whose capabilities stay largely constant compared to those of the other members. Staying constant does not mean that the state lacks ambitions, that it fails to secure further gains if presented with the opportunity, or that

Table 2.3 Rising powers, 1648–2018

Newcomers	Aspirants	Successors
Russia (1696–1721)	Prussia/Germany (1862–1890)	Great Britain (1688–1815)
Prussia (1713–1740)	Italy (1860–1943)	Imperial Germany (1890–1918)
Italy (1852–1860)	Japan (1931–1945)	Nazi Germany (1933–1945)
United States of America (1890–1898)	China (1950–1978)	Soviet Union (1945–1989)
Japan (1868–1905)	China (1978–2010)	China (2010–)
China (1947–1950)		

it forbids any gains or losses for any of the other great powers. But the net effect of such gains or losses on the existing distribution of power and status should be marginal. The result is that the state's existing power and status are consolidated, rather than being dramatically altered, either positively, as in the case of a rising power, or negatively, as in that of a declining power. Furthermore, status quo does not mean necessarily the same thing as being equal to the other great powers in power and status. If a state holds an advantage over the other great powers or conversely, it is at a disadvantage, but this situation does not change over time, this is compatible with the status quo in the sense of preserving things as they presently stand. Finally, status quo supposes risk aversion. As prospect theory holds, states, just like individuals, are risk-averse: they prefer smaller but certain gains to large but uncertain ones; at the same time they go to great lengths to prevent the possibility of loss. That means that status quo great powers are more preoccupied with what they might lose than with what they might gain.[52]

Status quo power may refer to two kinds of states: either the number one ("the dominant state") or to the rank-and-file members of the club ("the established powers"). As Samuel Huntington puts it, the number one "is able to exercise more influence on the behavior of more actors, with respect to more issues" than any other state can. In other words, this is the strongest state of the day, certainly the most capable due to its overall resources. From Westphalia to the present, there have been three such dominant actors—France, Great Britain, and the US.[53] Just how much stronger the number one must be relative to the other powers has historically varied. Hardly ever was there a state that controlled over half (50% plus) of the total capabilities in the system. There have been number one powers that on occasion controlled about a third of the total military power in the system (about 33% measured by army size); and the US even accounted for half the global economic production in the years after World War II. One could hypothesize, therefore, that a dominant state should control between a fourth and a third of the system's resources at a given time, and, more importantly, that it should surpass the rest of the great powers in overall performance across dimensions.

The sources of strength have also alternated across dominant states. France possessed the largest population, economy, army, and the best diplomatic service, but was surpassed by Holland in trade and, eventually, by a combination of Holland and Great Britain in naval power. Great Britain was the leader in industrial production, trade, finances, technology, naval power, and colonial empire, but was outmatched by other powers, principally Russia and France, in land forces. Only the US has succeeded in maintaining since 1945 a lead on the other great powers in every single dimension, although the nature of its preponderance has evolved. From 1945 to 1990, the US was unrivaled on economic terms, while in military matters it was ahead in some areas (navy, air force) and on par in others (nuclear weapons) with the Soviet Union (only ahead in ground forces). Since 1990, and, increasingly since the global recession of 2008, however, the US has faced sterner economic competition from China, while enjoying a disproportionate advantage in military and technology.[54]

Being more powerful than other states does not mean that the number one is all powerful, or a hegemon. Instead, it is a *primus inter pares*, first among equals. It cannot lay down the law to other states, acting as a government above the other great powers. However, following the rule of thumb that a regular great power can take on any other great power in a one-on-one conflict, dominant states are able to take on combinations of other (and potentially all) great powers even without great power allies. During the reign of Louis XIV, the Sun King, France fought several times to a standstill the combined forces of the other European powers. Indeed, Louis adopted as his motto "nec pluribus impar," translated as "a match for many." During the so-called continental system, a European-wide blockade and boycott put in place by Napoleon, a Great Britain on the edge of domination stood essentially alone against all the other great powers combined. As for the US, since it accounts in the post-Cold War for a larger percentage of the total global military spending than all the other great powers together (60% in 1998; 40% in 2018), post-Cold War that it could arguably engage any possible counter-combination of powers with a fair chance of success. Accordingly, this represents a strong deterrent to the creation of any anti-American coalition.[55]

Meanwhile, the established powers represent a group constituted of former rising powers that have gained successive admittance to the club, or former dominant states that have been relegated to the rank of ordinary club members. These rank-and-file members are neither in a position to become frontrunners, or in a position to be ejected from the club. This means that they do not contemplate either a substantial increase or a sharp drop in their share of power, and that they also do not anticipate either status promotion or demotion. While they may attempt to address the deficiencies in their capabilities portfolio or gain supremacy in those dimensions in which they are doing best, their concern is foremost to keep up with the Joneses, prioritizing "upkeep" over gains.

According to these criteria, the list of the status quo powers could be illustrated as shown in Table 2.4.[56]

Declining power is a term for a member of the great power club whose resources relative to those of another member or members decrease consistently,

Table 2.4 Status quo powers, 1648–2018

Dominant state	Established powers
France (1648–1763)	Holland (1648–1713)
Great Britain (1815–1918)	Sweden (1648–1679)
United States of America (1945–)	Spain (1648–1713)
	Austria (1648–1866)
	Russia (1721–1917)
	Prussia/Germany (1740–1862; 1945–)
	United States of America (1898–1945)
	Japan (1905–1931; 1945–)
	France (1870–)
	Great Britain (1945–)

and not merely temporarily. For decline to occur, falling behind in one component of power is not enough. As seen in the above, great powers often exhibit unbalanced portfolios of capabilities, meaning that they do better in some areas than others, while being able, however, to maintain their standing in the great power club. However, if a state falls behind the competition in all important areas, it would not be able to maintain its previous position.[57]

Unlike the case of the rising powers, decline is not just about the narrowing of the gap with the other powers, in the sense of the latter gaining upon club members without necessarily catching up or surpassing them. After all, even though a state may perform at worse levels than it used to in terms of overall economic performance or military spending, it may still do enough to keep it ahead of the competition. This may be the case of contemporary US, which is not performing as well as it did in previous decades compared to a rising China, yet still doing better in multiple domains by a comfortable margin.[58] Therefore, decline requires being overtaken by the competition in at least several power dimensions and being on the verge of being surpassed in all of them.[59]

Decline can be structural, political, or both. In the case of structural decline, a great power experiences reversals of fortune relative to other great powers in multiple material components of power, such as territory, demographics, wealth, military forces, or technology, that is to say the ability to innovate and implement innovations. Meanwhile, political decline, also referred to as decay or decadence, concerns a reduction of political and social power, hence a decrease in the ability of the state to mobilize, extract, and direct resources efficiently, which then results in diminishing economic productivity and setbacks in military contests. A great power may experience one form of decline without necessarily experiencing the other. It may, for instance, remain formidable in terms of its latent resources, while experiencing political dissension at the top, excessive consumption relative to investment, corruption and cronyism, civil unrest and violence, revolution, defeat in war, or even partial or total military occupation by another power. This was the case for Germany and Japan after World War II, and Holland after the War of the Spanish Succession. Conversely, a great power may slip behind in the club rankings due to objective power dynamics, such as slow growth in population, resource depletion, or economic stagnation, despite maintaining internal cohesion as well as a general record of success in military encounters (actually, it may be that its structural decline is helped along by its frequent involvement in costly, albeit victorious, wars and interventions abroad). This was the case of pre-revolutionary France, Spain, and Sweden after the Thirty Years' War, and of the Soviet Union in the 1980s. In the most serious instances, a great power experiences both structural and political decline at the same time, with one form affecting the other. Defeat in war may trigger a major loss of territory, revenue, and population; or the realization that the state has slipped up in major capabilities relative to other great powers sparks up internal political crisis.[60]

Declining powers differ from rising powers in that they follow opposite trajectories in the cycle of rising and falling powers: as the rising powers' fortunes improve, the declining powers' deteriorate. Declining powers also are distinct from status quo powers. First, their losses are severe, comprehensive, and lasting, not merely moderate, restricted to isolated dimensions, and provisional. Status quo powers can afford setbacks that they can make up for later on or through gains elsewhere without endangering their status. But declining powers face the danger of status demotion to the point of even losing their place in the great power club. These status losses may not be reversible. In fact, if the above great power identification criteria are accepted as valid, there has never been a great power that has managed to re-enter the club after being demoted, or a great power that has managed to regain its number one position after having lost it, although there have been cases of powers that have managed to regain their status as front-runners or established powers.[61] Second, again invoking prospect theory, while status quo powers prefer as well avoiding losses to making gains, declining powers are likely to be far more risk acceptant. The reason is that status quo powers only face limited setbacks or the prospect of loss. But declining powers have already experienced severe losses and are going therefore to be willing to go to great lengths to recoup them, just as a gambler at a horse race runs high risks on the last race of the day in the hope to regain all his or her previous losses.[62]

Declining power may indicate several kinds of states. The first type comprises declining number ones, who are on the brink of losing their supremacy as the strongest state in the system, and risk being demoted from the number one spot into the pool of rank-and-file club members. This has been the case of both France and Britain. The second type is that of declining established powers that are on the brink of losing their status as members of the club altogether, thus being reduced to average states. Every state that has been eventually driven out of the club has thus gone through an interval of decline. The third type is that of resurgent powers, which are not actually rising powers, but rather declining states attempting to make a comeback. A possible fourth type may consist of Imperial China and the Ottoman Empire, who did not hold any status in the club, yet had to abandon their claim of being the world leader, and thus lost commensurable power relative to the members of the great power club and status in the regions they controlled.

Hence, the list of declining great powers could be illustrated as shown in Table 2.5.[63]

Table 2.5 Declining powers, 1648–2018

Declining dominant state	Declining established powers	Resurgent powers	Declining quasi-powers
France (1763–1815)	Sweden (1679–1721)	Spain (1713–1763)	Ottoman Empire (1774–1918)
Great Britain (1918–1945)	Holland (1713–1740)	France (1815–1870)	Imperial China (1841–1912)
	Austria-Hungary (1866–1918)	Soviet Union (1917–1945) Russia (1991–)	

Table 2.6 Evolution of great powers in the club, 1648–2018

Great Power	Rising Power	Status Quo Power	Declining Power
France		1648–1763; 1870-	1763–1870
England/Great Britain	1688–1815	1815–1918; 1945-	1918–1945
Austria/Austria-Hungary		1648–1866	1866–1918
Spain		1648–1713	1713–1763
Sweden		1648–1679	1679–1721
Holland		1648–1713	1713–1748
Russia/Soviet Union	1697–1721; 1945–1991	1721–1917	1917–1945; 1991–
Prussia/Germany	1713–1740; 1862–1945	1740–1862; 1945–	
Italy	1852–1943		
United States of America	1890–1898	1898–	
Japan	1868–1945	1945–	
China	1947–		

Summing up the results for the states composing the great power club for the interval running from 1648 to the present, which are illustrated in Table 2.1, one arrives at the evolution shown in Table 2.6 for each great power.

Having discussed the criteria for identification of great powers, the composition of the great power club since its founding to the present, as well the classification of great powers in rising, status quo, and declining powers, the book will now turn to the question of the grand strategies characteristic of each type.

Notes

1 Levy, *War in the Modern Great Power System*, 10–18; Buzan, *United States and the Great Powers*, 47–71; Lebow, *Why Nations Fight*; Wight, *Power Politics*, 41–53; Bull, *Anarchical Society*, 194–220; Danilovic, *When Stakes Are High*, 28–47; Sarkees and Wayman, *Resort to War*, 34–6; Modelski, *Principles of World Politics*.

2 Each large indicator is composed in turn of a host of sub-components. The most detailed power analysis to date identifies no fewer than 40 separate smaller elements, considered "the minimal necessary quantitative information for judging national capabilities in the post-industrial age." For instance, military capability alone is broken down into absolute size of defense budget; size of defense budget relative to GDP and peers; education level of enlisted and officers; number of advanced training facilities; holdings of high-leverage combat systems; number of research, development, test, and evaluation facilities; extent of training abroad; number of high-level training exercises; as well as various indicators of technology and integration—no fewer than nine elements. Furthermore, this analysis is confined to material power exclusively, without even taking into account any non-material dimensions, whether social, political, or diplomatic. Tellis, Bially, Layne, and McPherson, *Measuring National Power*, 179–82.

3 Central Intelligence Agency, *CIA World Factbook 2018* at https://www.cia.gov/libra ry/publications/the-world-factbook/rankorder/2001rank.html; Stockholm International Peace Research Institute, Trends in World Military Expenditures, 2017 at https:// www.sipri.org/sites/default/files/2018-05/sipri_fs_1805_milex_2017.pdf;   Global Innovation Index, 2017 Report at https://www.globalinnovationindex.org/analysis-ind icator; Correlates of War, "National Material Capabilities v. 5," at http://www.correlate sofwar.org/data-sets/national-material-capabilities.

4 Brooks and Wohlforth, "Rise and Fall"; Waltz, *Theory of International Politics*, 127–31.

5 Gilpin, *War and Change in World Politics*, 23–4; Wohlforth, *Elusive Balance*, 26–8.

6 Brooks and Wohlforth, "Rise and Fall," 21–32; Beckley, "China's Century?"; Managi and Kumar, eds. *Inclusive Wealth Report*; Coyle, *GDP*.

7 Johnston, *Cultural Realism*; Berger, "Norms, Identity and National Security," 317–56; Kennedy, *Rise and Fall of British Naval Mastery*; Crawl, "Alfred Thayer Mahan," 444–80; Hardesty and Eisman, *Epic Rivalry*; Siddiqi, *Challenge to Apollo*.

8 James Fearon, "What Is Identity (As We Now Use the World)?" unpublished manu-script available at https://web.stanford.edu/group/fearon-research/cgi-bin/wordpress/ wp-content/uploads/2013/10/What-is-Identity-as-we-now-use-the-word-.pdf.

9 Dahl, *Modern Political Analysis*, 27.

10 Buzan, *United States and Great Powers*, 47; Buzan, *People, States, and* Fear; Buzan and Waever, *Regions and World Powers*; Posen, "Command of the Commons."

11 Levy, *War in the Modern Great Power System*, 16–17.

12 Bull, *Anarchical Society*, 53–71, chap. 9.

13 Wright, *Study of War*, 220–2; Levy, *War in the Modern Great Power System*, 3, 97–9.

14 von Ranke, "Great Powers," 202–3.

15 Levy, *War in the Modern Great Power System*, 10–3; Wight, *Power Politics*, 18, 46–8, 52–3; Taylor, *Struggle for Mastery in Europe*, XXIV; Modelski, *Principles*, 150.

16 Morgenthau, *Politics Among Nations*, 79.

17 Quoted in Schlessinger, *Robert Kennedy and His Times*, 705.

18 Baldwin, "Success and Failure in Foreign Policy"; Jervis, *Meaning of Nuclear Revolution*, 186–91; Johnson and Tierney, *Failing to Win*.

19 Levy, *War in the Modern Great Power System*, 17.

20 Elrod, "Concert of Europe," 163–6; Gulick, *Europe's Classical Balance of Power*, 70–2; Schroder, *Transformation*, 6–7; Jervis, "From Balance to Concert."

21 One telltale sign one is dealing with a great power is the degree of attention paid to it by other states, as measured by means of the number of foreign diplomatic envoys permanently stationed in its capital. Renshon, *Fighting for Status*; Singer and Small, "Composition and Status Ordering of the International System."

22 Buzan, *United States and Great Powers*, 67, 70–1.

23 Levy, *War in the Modern Great Power System*, 17.

24 Gilpin, *War and Change*, 30, 34; Tajfel and Turner, "Social Identity Theory of Intergroup Behavior"; Johnston, *Social States*, 96–9. For a critique see Mercer, "Illusion of International Prestige."

25 O'Neill, *Honor, Symbols and War*, 108–12, 146–52; also see Lindner, *Making Enemies*; Miller, *Humiliation*; Neu, *Sticks and Stones*.

26 Elrod, "Concert of Europe," 167; Schroder, *Transformation*, 7–8; Gulick, *Europe's Classical Balance of Power*, 75–6.

27 O'Neill, *Honor*, 89, 143; Kissinger, *White House Years*, 228; Lebow, *Why Nations Fight*, 171–2.

28 Renshon, "Losing Face and Sinking Costs"; Lebow, *Why Nations Fight*, 171–2.

29 Quoted in Gulick, *Europe's Classical Balance of Power*, 28–9.

30 An interesting question involves whether China (the Popular Republic of China) from 1949 to 1971, the year it assumed its permanent seat in the UN Security Council, should

be seen as a great power. On the one hand, it behaved as a great power and had great power capabilities. On the other hand, it was formally denied recognition not only as a great power, but also as a state. One could argue that China was a great power based on its capabilities alone whether others chose or not to acknowledge it, thus discounting the role of recognition. Yet, it becomes clear that even those great powers that rejected any formal legitimacy to the government in Beijing, admitted in practice that it was a great power by taking it into account in their strategic calculations and by being careful not to humiliate it. Chang, *Friends and Enemies*.

31 The exception to the rule is Italy, the weakest great power, and, after Sweden, the state with the shortest tenure in the great power club. Italy was much disparaged behind closed doors by other great powers for not meeting the requisite great power criteria, which led to a constant Italian effort to demonstrate its credentials. However, Italy was included in the governance of the international system at major conferences, and its grand strategy in both world wars was a major factor affecting the decisions of the other great powers. Mack-Smith, *Modern Italy*; Bosworth, *Italy*.

32 Levy, *War in the Modern Great Power System*, 21–4. Other scholars supporting this view are Dehio, *Precarious Balance*; Mattingly, *Renaissance Diplomacy*; Modelski, *Principles of World Politics*. For the Congress of Vienna see Webster, *Congress of Vienna*.

33 Osiander, "Sovereignty"; Nexon, *Struggle for Power*, 273–81.

34 Levy, *War in the Modern Great Power System*, 74–5; Petrie, *Early Diplomatic History*.

35 Croxton and Parker, "'A Swift and Sure Peace,'" 71–100, 72, 97.

36 Aron, *Peace and War*, 94; Bull and Watson, eds. *Expansion of the International Society*, 1; Jervis, *System Effects*, 6.

37 Several scholars trace the origins of the great powers to the the mid-seventeenth to eighteenth century interval. McKay and Scott, *Rise of Great Powers*; Wolf, *Emergence of Great Powers*; Hinsley, *Power and the Pursuit of Peace* 153, 160–3; Craig and George, *Force and Statecraft*; Lebow, *Why Nations Fight*.

38 This is no accident. Many of the histories of the great powers' emergence turn up at a closer look to be diplomatic histories, concerned primarily with the origins and development of diplomatic relations. Petrie, *Early Diplomatic History*; Mattingly, *Renaissance Diplomacy*; Mowat, *History of European Diplomacy*.

39 Osiander, "Sovereignty," 270–2, 278–80; Wolf, *Emergence*, 5; Mowat, *History of European Diplomacy*, 28–9, 60–1; Parker, *Military Revolution*, 1–2, 18, 23–4.

40 Gilbert, *To the Farewell Address*, 89–104; Meinecke, *Machiavellians*; Hinsley, *Power and the Pursuit of Peace*.

41 Levy, *War in the Modern Great Power System*, 25–6; Sarkees and Wayman, *Resort to War*, 35. An additional problem for the COW dataset is that it fails to specify the exact criteria for readmission in the club. Germany and Japan, while excluded from the club from 1945 onwards, are readmitted in 1991, despite, especially in the case of the latter, any perceptible change in the late 1980s in their foreign policy or the treatment received from other states.

42 What varies the most among scholars is not the membership of the club, but the length of tenure of individual great powers. With a few notable differences, the dates are derived from Levy, *War in the Modern Great Power System*, 37–44. The differences are as follows. The origins of the great power system are traced to 1648 rather than to 1494. Spain's great power tenure ends with its defeat in the Seven Years' War, rather than with occupation by France in 1808, the reason being that Spain could not make a successful comeback after the 1760s, being unable to fight another great power on its own, and regularly meeting with defeat on the battlefield. Meanwhile, Holland's tenure as a great power is extended to 1748, rather than ending abruptly in 1713. In effect, 1713 does not make sense as an exit date as Holland had been one of the victors of the War of Spanish Succession and one of the beneficiaries of the resulting Utrecht Peace Treaty. In the

1720s, Holland was still a commercial, technological, naval, and colonial powerhouse. It was its lack of reaction in the face of French invasion in defending the Austrian Netherlands (Belgium), an area previously deemed crucial to its security and for which it had fought fiercely against Louis XIV, coupled with the decline of its military force and economic power that signaled that it no longer counted as a great power. Japan is considered as an uninterrupted member of the great power club from 1905 onwards, instead of being excluded following defeat and occupation in World War II. Following the 25 years recovery rule, by the 1970s, Japan was experiencing an economic boom and, despite the limitation of military expenses, its Self Defense Forces were on par in size with troops maintained by France and Great Britain. Finally, as detailed later on, the Ottoman Empire having never really been a fully-fledged member of the club, could not have lost its position as a result of defeat, as Levy argues, in 1699. Kamen, *Empire*; Lynch, *Bourbon Spain*; Israel, *Dutch Republic*; Danilovic, *When Stakes Are High*, 42–4.

43 Singer and Small, *Wages of War*; Modelski, *Principles*, 144–5; Levy, *War in the Modern Great Power System*, 47–9; Kennedy, *Rise and Fall*; Danilovic, *When Stakes Are High*, 29–45; Buzan, *US and World Powers*, 69–71; Sarkees and Wayman, *Resort to War*, 35; Lebow, *Why Nations Fight*, 103. Only Buzan fails to include Italy and Austria as great powers at the end of the nineteenth century, and considers adding the EU.

44 Kennedy, *Rise and Fall*, 149, 147–50; Correlates of War, "National Material Capabilities."

45 Levy, *War*, 35–7. However, if the 25 years' recovery rule holds, the Ottoman Empire continued to be a great power for almost another three-quarters of a century after Carlowitz. The Ottomans managed to recover in the eighteenth century a good deal of their territorial losses, notably Morea (the Peloponnese), and Serbia, including Belgrade; and even to make gains in the Caucasus (Georgia, Azerbaijan). Significantly, in 1711, they inflicted a comprehensive defeat on Peter the Great's Russia and from 1736 to 1739 they fought both Austria and Russia to a standstill. A better date for the supposed exit from the club of the Ottoman Empire is the Küçük Kaynarca Treaty of 1774, which represented a turning point, because the Porte was giving up for the first time territory inhabited by Muslims (the Crimea to Russia). There was no recovery after the treaty, but an uninterrupted succession of humiliating defeats, provincial rebellions, and foreign interference. As scholarship points out, the Ottoman Empire was never again able to fight a great power without support. Thus, Küçük Kaynarca is the origin date of the so-called Eastern Question—the seemingly inevitable collapse of the Ottoman Empire and the division of its provinces. Shaw, *History of Ottoman Empire and Modern Turkey*, 224, 229–30, 244–5, 248–50; Neumann, "Political and Diplomatic Developments," 44–62, esp. 57–8.

46 Bull and Watson, *Expansion of International Society*, 143–69, 171–83; Neumann, "Status Is Cultural," 85–112; Neumann, *Uses of the Other*, 43–54; Bull, *Anarchical Society*, 13–4; Gulick, *Europe's Classical Balance of Power*, 14–6; Hinsley, *Power and the Pursuit of Peace*, 14–9, 30–4.

47 Mote, *Imperial China*; Fairbank, *Chinese World Order: Traditional China Foreign*; Bull and Watson, *Expansion of International Society*, 143; Shaw, *Between Old and New*.

48 Lebow, *Why Nations Fight*, 97–104, with the proviso that Lebow substitutes great powers for status quo powers. The wider literature recognizes a basic distinction between status quo and revisionist states. Yet, as Johnston observes, the definitions of both are not only vague, but also undertheorized. Johnston, "Is China a Status Quo Power," 8. Also see Schweller, "Managing the Rise of Great Powers"; Davidson, *Origins of Revisionist and Status Quo States*; Morgenthau, *Politics Among Nations*; Wolfers, *Discord and Collaboration*.

49 Chestnut and Johnston, "Is China Rising," 237–59.

50 The interval in which a state is a rising power ends either with the achievement or defeat of its bid at improving its standing in international rankings. The start of the interval usually signals the coming to power of decisionmakers—Peter the Great (Russia); Frederick William I (Prussia); Cavour (Italy); the Meiji Oligarchs (Japan); Benjamin Harrison (the US); Bismarck (Prussia); Mussolini (Italy); Deng Xiaoping (China); William of Orange (Great Britain); Kaiser Wilhelm II (Germany); and Hitler (Germany) who adopted measures aimed to claim a significant improvement in status and international governance for the state. The exceptions are those (China, Japan, the Soviet Union) whose leaders had been in charge already for years or even decades, but began expressing claims for a substantially larger share of status and governance because of changing circumstances usually in the aftermath of a war.

51 Dominic Wilson and Rupa Purushothoman, "Dreaming with the BRICs: The Path to 2050," *Global Economic Paper* 99, available at http://www.goldmansachs.com/ our-thinking/topics/brics/brics-reports-pdfs/brics-dream.pdf; Jim O'Neill and Anna Stupnytska, "The Long Term Outlook for the BRICs and N–11 Post Crisis," *Global Economics Paper* 192 available at http://www2.goldmansachs.com/ideas/brics/long -term-outlook-doc.pdf; Asa Johansson et al., "Looking to 2060: Long Term Global Growth Prospects," *OECD Economic Policy Papers 3*, at http://www.oecd.org/eco /economicoutlookanalysisandforecasts/2060%20policy%20paper%20FINAL.pdf; National Intelligence Council, "Global Trends 2025: A Transformed World," 28–37, at http://www.aicpa.org/research/cpahorizons2025/globalforces/downloadabledocu ments/globaltrends.pdf

52 Morgenthau, *Politics Among Nations*, 46–7; Levy, "Loss Aversion," 193–221; Tversky and Kahneman, "Prospect Theory." Also see Johnston, "Is China a Status Quo Power"; Schweller, "Bandwagoning for Profit," 104–5; Wolfers, *Discord and Collaboration*, 92, 96–7.

53 Huntington, "Why International Primacy Matters," 68; Buzan, *United States and Great Powers*, 68–71; Lebow, *Why Nations Fight*, 103, 99–100; Wight, *Power Politics*, 30–40.

54 Thompson and Levy, "Hegemonic Threats and Great Power Balancing in Europe," 20; Wohlforth, "Stability of a Unipolar World," 12–16; Niou, Ordeshook, and Rose, *Balance of Power*, 76.

55 Wight, *Power Politics*, 34; Brooks and Wohlforth, *World Out of Balance*.

56 The interval in which a state starts being an established power usually coincides with its induction in the great power club. This is true of all new great powers with the exceptions of Italy and China, whose admission into the great power club was in the case of the former questionable, and in the second formally contested, and who, as a consequence, sought to rise further in order to put all doubts about their credentials to rest. Other exceptions are former dominant states, such as France and Britain, who turn to status quo once they accept that their loss of status is irreversible. For dominant states, the start of the status quo interval is marked by victory in a war against their main competitor for supremacy in the system. The end of the status quo interval is traced from an event such as defeat in war (France, Austria, Russia) or a substantial change of traditional policy (such as Britain's abandonment of the Two-Power Standard), which signals the elites' acceptance that the state has entered a period of decline.

57 Wohlforth, "How Not to Evaluate Theories." For a recent generic argument that decline means being overtaken in only a single dimension, GDP, see MacDonald and Parent, *Twilight of the Titans*.

58 Brooks and Wohlforth, "Rise and Fall"; Brooks and Wohlforth, *America Abroad: The United States' Global Role in the Twenty-First Century* (New York: Oxford University Press, 2016); Beckley, "China's Century?"

59 Power transition theory's latest version holds that a declining power is more likely to engage in war against a challenger when the latter approaches its power levels (power

parity). This reverses its former position that war is most likely after the challenger overtakes the declining power. Tammen et al., eds., *Power Transitions*; DiCicco and Levy, "Power Transition Research Program," 109–58.

60 Nye, *Bound to Lead*, 14–16. Raymond Aron suggests a dichotomy between decadence (as a value judgment concerning the state losing its vitality or its capacity for collective action) and decline (as a structural power relation). Aron, *In Defense of Decadent Europe*, xv–xvii. For the causes of structural decline, see Gilpin, *War and Change*, 159–85.

61 The most notable examples being France after 1815 and the Soviet Union after 1917. China might be mentioned as the lone example of a state that successfully regained its great power status, except that, if Imperial China was never considered a member of the club to start with, this would not be re-admission.

62 Levy, "Loss Aversion."

63 Lebow, *Why Nations Fight*, 102–3. The start of decline is traced from the realization by the central decision-makers that the state is experiencing decline rather than from the objective decreasing trend in capabilities. Thus, the onset of decline frequently coincides with the experience of defeat, weakening, or territorial loss that brings about reality to bear on the elites. The end of decline is marked in the case of dominant and established powers by the loss of their former status, or in the case of resurgent powers by either recovery or by a clear defeat of their efforts at recouping losses. Gilpin, *War and Change*; Friedberg, *Weary Titan*.

Bibliography

Aron, Raymond. *In Defense of Decadent Europe*. South Bend: Regnery/Gateway, 1979.

Aron, Raymond. *Peace and War: A Theory of International Relations*. London: Weidenfeld and Nicolson, 1966.

Baldwin, David. "Success and Failure in Foreign Policy." *Annual Review of Political Science* 3 (2000): 167–82.

Beckley, Michael. "China's Century? Why America's Edge Will Endure." *International Security* 36 (Winter 2011): 41–78.

Berger, Thomas. "Norms, Identity and National Security in Germany and Japan." In *The Culture of National Security*, edited by Peter Katzenstein. New York: Columbia University Press, 1996.

Bosworth, RJB. *Italy, the Least of the Great Powers: Italian Foreign Policy Before the First World War*. New York: Cambridge University Press, 1979.

Brooks, Stephen and William Wohlforth. *America Abroad: The United States' Global Role in the Twenty-First Century*. New York: Oxford University Press, 2016.

Brooks, Stephen and William Wohlforth. "The Rise and Fall of the Great Powers in the Twenty-First Century." *International Security* 40 (Winter 2015): 7–53.

Brooks, Stephen and William Wohlforth. *World Out of Balance: International Relations and the Challenge of American Primacy*. Princeton: Princeton University Press, 2008.

Bull, Hedley. *Anarchical Society: A Study of Order in World Politics*. New York: Columbia University Press, 1977.

Bull, Hedley and Adam Watson, eds. *The Expansion of the International Society*. Oxford: Clarendon Press, 1984.

Buzan, Barry. *People, States, and Fear: The National Security Problem in International Relations*. Chapel Hill: University of North Carolina Press, 1983.

Buzan, Barry. *The United States and the Great Powers: World Politics in the Twenty-First Century*. New York: Polity, 2004.

Buzan, Barry and Ole Waever. *Regions and World Powers*. New York: Cambridge University Press, 2003.

Chang, Gordon. *Friends and Enemies: The United States, China and the Soviet Union, 1948–1972*. Stanford: Stanford University Press, 1990.

Chestnut, Sheena and Alastair Iain Johnston. "Is China Rising?" In *Global Giant: Is China Changing the Rules of the Game?*, edited by Eva Paus, Penelope Prime, and Jon Western. New York: Palgrave Macmillan, 2009.

Coyle, Dana. *GDP: A Brief but Affectionate History*. Princeton: Princeton University Press, 2014.

Craig, Gordon and Alexander George. *Force and Statecraft: Diplomatic Problems of Our Time*, 3rd edn. New York: Oxford University Press, 1995.

Crawl, Philip. "Alfred Thayer Mahan: The Naval Historian." In *Makers of Modern Strategy: From Machiavelli to the Nuclear Age*, edited by Peter Paret. Princeton: Princeton University Press, 1986.

Croxton, Derek and Geffrey Parker. "'A Swift and Sure Peace': The Congress of Westphalia, 1643–1648." In *The Meaning of Peace: Rulers, States, and the Aftermath of Wars*, edited by Williamson Murray and James Lacey. New York: Cambridge University Press, 2008.

Dahl, Robert. *Modern Political Analysis*. Englewood Cliffs: Prentice Hall, 1984.

Danilovic, Vesna. *When the Stakes Are High: Deterrence and Conflict Among Major Powers*. Ann Arbor: University of Michigan Press, 2002.

Davidson, Jason. *The Origins of Revisionist and Status Quo States*. New York: Palgrave Macmillan, 2006.

Dehio, Ludwig. *The Precarious Balance: Four Centuries of the European Power Struggle*. New York: Knopf, 1962.

DiCicco, Jonathan and Jack Levy. "The Power Transition Research Program." In *Progress in International Relations Theory: Appraising the Field*, edited by Colin Elman and Miriam Fendius Elman. Cambridge: MIT Press, 2003.

Elrod, Robert. "The Concert of Europe: A Fresh Look at an International System." *World Politics* 28 (January 1976): 159–74.

Fairbank, John King. *The Chinese World Order: Traditional China Foreign Relations*. Cambridge: Harvard University Press, 1968.

Friedberg, Aaron. *Weary Titan: Britain and the Experience of Relative Decline, 1895–1905*. Princeton: Princeton University Press, 1988.

Gilbert, Felix. *To the Farewell Address: Ideas of Early American Foreign Policy*. Princeton: Princeton University Press, 1961.

Gilpin, Robert. *War and Change in World Politics*. New York: Cambridge University Press, 1981.

Gulick, Edward Vose. *Europe's Classical Balance of Power*. Ithaca: Cornell University Press, 1955.

Hardesty, Von and Gene Eisman. *Epic Rivalry: The Inside Story of the Soviet and American Space Race*. Washington, DC: National Geographic, 2007.

Hinsley, FH. *Power and the Pursuit of Peace: Theory and Practice in the History of Relations Between States*. Cambridge: Cambridge University Press, 1963.

Huntington, Samuel. "Why International Primacy Matters." *International Security* 17 (Spring 1993): 68–83.

Israel, Jonathan. *The Dutch Republic: Its Rise, Greatness, and Fall, 1477–1806*. Oxford: Oxford University Press, 1995.

Jervis, Robert. "From Balance to Concert: A Study of International Security." *World Politics* 35 (October 1985): 58–79.

Jervis, Robert. *System Effects: Complexity in Political and Social Life.* Princeton: Princeton University Press, 1997.

Jervis, Robert. *The Meaning of the Nuclear Revolution: Statecraft and the Prospect of Armageddon.* Ithaca: Cornell University Press, 1989.

Johnson, Dominic and Dominic Tierney. *Failing to Win: Perceptions of Victory and Defeat in International Politics.* Cambridge: Harvard University Press, 2006.

Johnston, Alastair Iain. *Cultural Realism: Strategic Culture and Grand Strategy in Chinese History.* Princeton: Princeton University Press, 1995.

Johnston, Alastair Iain. "Is China a Status Quo Power?" *International Security* 27 (Spring 2003): 5–56.

Johnston, Alastair Iain. *Social States: China in International Institutions, 1980–2000.* Princeton: Princeton University Press, 2000.

Kamen, Henry. *Empire: How Spain Became a World Power: 1492–1763.* New York: Harper Collins, 2003.

Kennedy, Paul. *The Rise and Fall of British Naval Mastery.* London: Allen Lane, 1976.

Kissinger, Henry. *White House Years.* Boston: Little Brown, 1979.

Lebow, Richard Ned. *Why Nations Fight: Past and Future Motives for War.* Cambridge: Cambridge University Press, 2010.

Levy, Jack. "Loss Aversion, Framing Effects, and International Conflict." In *Handbook of War Studies II,* edited by Manus Midlarsky. Ann Arbor: University of Michigan Press, 2000.

Levy, Jack. *War in the Modern Great Power System, 1495–1975.* Lexington: University Press of Kentucky, 1983.

Lindner, Evelyn. *Making Enemies: Humiliation and International Conflict.* Westport: Praeger, 2006.

Lynch, John. *Bourbon Spain, 1700–1808.* Oxford: Blackwell, 1989.

MacDonald, Paul and Joseph Parent. *Twilight of the Titans: Great Power Decline and Retrenchment.* Ithaca: Cornell University Press, 2018.

Mack-Smith, Denis. *Modern Italy: A Political History.* Ann Arbor: University of Michigan Press, 1997.

Managi, Shunsuke and Pushpam Kumar, eds. *Inclusive Wealth Report: Measuring Progress Towards Sustainability.* New York: Routledge, 2018.

Mattingly, Garrett. *Renaissance Diplomacy.* London: Cape, 1955.

McKay, Derek and HM Scott. *The Rise of the Great Powers, 1648–1815.* New York: Longman, 1983.

Meinecke, Friedrich. *The Machiavellians: The Doctrine of Raison d'État and Its Place in Modern History.* New Haven: Yale University Press, 1957.

Mercer, Jonathan. "The Illusion of International Prestige." *International Security* 41 (Spring 2017): 133–68.

Miller, William Ian. *Humiliation and Other Essays on Honor, Social Discomfort, and Violence.* Ithaca: Cornell University Press, 1993.

Modelski, George. *Principles of World Politics.* New York: Free Press, 1972.

Morgenthau, Hans. *Politics Among Nations: The Struggle for Power and Peace.* New York: Knopf, 1979.

Mote, Frederick. *Imperial China, 900–1800.* Cambridge: Harvard University Press, 1999.

Mowat, RB. *A History of European Diplomacy: 1451–1789.* New York: Longmans, 1928.

Neu, Jerome. *Sticks and Stones: The Philosophy of Insults*. New York: Oxford University Press, 2008.

Neumann, Christoph. "Political and Diplomatic Developments." In *The Cambridge History of Turkey, vol. 3 The Later Ottoman Empire*, edited by Suraya Faroqhi, 1603–1839. New York: Cambridge University Press, 2006.

Neumann, Iver. "Status Is Cultural: Durkheimian Poles and Weberian Russians Seek Great Power Status." In *Status in World Politics*, edited by Deborah Larson, TV Paul, and William Wohlforth. New York: Oxford University Press, 2014.

Neumann, Iver. *Uses of the Other: 'The East' in European Identity Formation*. Minneapolis: University of Minnesota Press, 1999.

Nexon, Daniel. *The Struggle for Power in Early Modern Europe: Religious Conflict, Dynastic Empires, and International Change*. Princeton: Princeton University Press, 2009.

Niou, Emerson, Peter Ordeshook, and Gregory Rose. *The Balance of Power: Stability in International Systems*. New York: Cambridge University Press, 1989.

Nye, Joseph. *Bound to Lead: The Changing Nature of American Power*. New York: Basic Books, 1990.

O'Neill, Barry. *Honor, Symbols and War*. Ann Arbor: Michigan University Press, 1999.

Osiander, Andreas. "Sovereignty, International Relations, and the Westphalian Myth." *International Organization* 55 (Spring 2001): 251–87.

Parker, Geoffrey. *The Military Revolution: Military Innovation and the Rise of the West, 1500–1800*. Cambridge: Cambridge University Press, 1988.

Petrie, Charles. *Early Diplomatic History*. New York: Macmillan, 1949.

Posen, Barry. "Command of the Commons: The Military Foundations of US Supremacy." *International Security* 28 (Summer 2003): 5–46.

Ranke, Leopold von. "The Great Powers." In *Leopold Ranke: The Formative Years*, edited by Theodore von Laue. Princeton: Princeton University Press, 1950.

Renshon, Jonathan. *Fighting for Status: Hierarchy and Conflict in World Politics*. Princeton: Princeton University Press, 2017.

Renshon, Jonathan. "Losing Face and Sinking Costs: Experimental Evidence on the Judgment of Political and Military Leaders." *International Organization* 69 (June 2015): 659–95.

Sarkees, Meredith Reid and Frank Wayman. *Resort to War: A Data Guide to Inter-state, Extra-state, Intra-State, and Non-State Wars, 1816–2007*. Washington, DC: CQ Press, 2010.

Schlessinger, Arthur. *Robert Kennedy and His Times*. Boston: Houghton Mifflin, 1978.

Schroder, Paul. *The Transformation of European Politics, 1763–1848*. New York: Oxford University Press, 1994.

Schweller, Randall. "Bandwagoning for Profit: Bringing the Revisionist State Back In." *International Security* 19 (Summer 1994): 72–107.

Schweller, Randall. "Managing the Rise of Great Powers: History and Theory." In *Engaging China: The Management of an Emerging Power*, edited by Alastair Iain Johnston and Robert Ross. New York: Routledge, 1999.

Shaw, Stanford. *Between Old and New: The Ottoman Empire Under Sultan Selim III, 1789–1907*. Cambridge: Harvard University Press, 1971.

Shaw, Stanford. *History of the Ottoman Empire and Modern Turkey*, Vol. 1. New York: Cambridge University Press, 1976.

Siddiqi, Asif. *Challenge to Apollo: The Soviet Union and the Space Race, 1945–1974*. Washington, DC: National Aeronautics and Space Administration, NASA History Div., Office of Policy and Plans, 2000.

Singer, David and Melvin Small. "The Composition and Status Ordering of the International System." *World Politics* 18 (January 1966): 236–82.

Singer, David and Melvin Small. *The Wages of War, 1816–1965: A Statistical Handbook.* New York: John Wiley and Sons, 1972.

Tajfel, Henry and John Turner. "The Social Identity Theory of Intergroup Behavior." In *Psychology of Intergroup Relations*, edited by Stephen Worchel and William Austin. Chicago: Nelson-Hall, 1986.

Tammen, Ronald, Douglas Lemke, Carole Alsharabati, Brian Efird, Jacek Kugler, Allan Stam III, Mark Andrew Abdollahian, and AFK Organski, eds. *Power Transitions: Strategies for the 21st Century.* New York: Chatham House Publishers, 2000.

Taylor, AJP. *The Struggle for Mastery in Europe, 1848–1918.* Oxford: Clarendon, 1954.

Tellis, Ashley, Janice Bially, Christopher Layne, and Mellissa McPherson. *Measuring National Power in the Postindustrial Age.* Santa Monica: Rand, 2003.

Thompson, William and Jack Levy. "Hegemonic Threats and Great Power Balancing in Europe, 1495–1999." *Security Studies* 14 (January–March 2005): 1–33.

Tversky, Amos and Daniel Kahneman. "Prospect Theory: An Examination of Decisions under Risk." *Econometrica* 47, no. 2 (March 1979): 263–91.

Waltz, Kenneth. *Theory of International Politics.* Reading: Addison Wesley, 1979.

Webster, CK. *The Congress of Vienna, 1814–1815.* London: G. Bell & Sons, 1950.

Wight, Martin. *Power Politics.* New York: Holmes and Meier, 1973.

Wohlforth, William. "How Not to Evaluate Theories." *International Studies Quarterly* 56 (2012): 219–22.

Wohlforth, William. *The Elusive Balance: Power and Perceptions During the Cold War.* Ithaca: Cornell University Press, 1993.

Wohlforth, William. "The Stability of a Unipolar World." *International Security* 24 (Summer 1999): 5–41.

Wolf, John B. *The Emergence of the Great Powers, 1685–1715.* New York: Harper & Row, 1951.

Wolfers, Arnold. *Discord and Collaboration: Essays on International Politics.* Baltimore: Johns Hopkins Press, 1962.

Wright, Quincy. *A Study of War.* Chicago: University of Chicago Press, 1965.

3 The grand strategies of rising powers

This chapter discusses several of the most significant grand strategies available to rising powers. It starts by presenting the key points of each grand strategy, the conditions under which it is likely to work best, as well as its limitations and drawbacks. The chapter, then, proceeds to showcase each of the grand strategies through an illustrative historical example of a rising power.

Self-strengthening

Self-strengthening refers to a grand strategy of catching up to better-off great powers through fundamental reforms aimed at reshaping the state on the model of the most advanced states of the day. This is best seen in the case of Peter the Great's Russia and of Meiji Japan. Reforms are not restricted to importing new technology, weapons, military tactics, and capital from abroad. As the historian Geoffrey Parker writes, for lasting success to be achieved, "simply copying weapons picked on the battlefield would never suffice, because doing so, required the 'replication' of the whole social and economic system that underpinned the capacity [of the model state] to innovate and respond swiftly."[1] To this extent, self-strengthening is more than just increasing one's material capabilities by emulating the successful practices of the great powers, a tactic referred to as internal balancing. A successful state should be able to do more than perennially follow suit in the wake of others. Eventually, it needs to strive ahead by championing its brand of new technologies, practices, and institutions. Furthermore, it also needs to respond efficiently and in a timely manner to crises, which requires the ability to improvise on the spot, instead of copying what others have done. Failure in both these regards means that no matter how good an imitator a state is, it will be always lag behind those powers that it imitates.[2] Hence, a rising power wishing to compete with the best needs to reproduce the wider foundations on which their economic prosperity and military success are built, which requires a revolution in its administration, education, culture, economy, infrastructure, institutions, and even social customs.

Reforms are carried out by foreign advisors or/and by leaders educated abroad and, as a result, require continual close cooperation with foreign "benefactor" states. In this sense, self-strengthening is antithetical to autarchy or self-sufficiency, in which the state restricts contacts with foreign actors, as well as their influence in its territory, so as to maximize its independence. By contrast to autarchy, self-strengthening recognizes that the best results can only be achieved by increasing the exposure of the state to foreign ideas and practices. However, over the long run self-strengthening seeks to reduce the state's initial dependence on foreign aid, by developing home-grown variants of advanced economic enterprises, military forces, or institutions.

Self-strengthening tackles the problem of potential great power opposition to the state's rise by ensuring a fast and concomitant development of its economic and military power, with both dimensions being judged as comparable in importance, seen for instance in Meiji Japan's slogan *fukoku kyōhei* (rich country, strong military). The goal is to have, available as soon as possible, sufficient forces to play on the same military level with a great power opponent. Thus, military development is not to be sacrificed to the larger cause of social, economic, and cultural transformation, by being tackled last, once the state would dispose of sufficient resources to afford it. Instead, it is part and parcel of the transformation agenda from the moment reforms are launched, so that the military power of the state grows in step with the other changes taking place. The new strengthened military is able both to protect the state while it is completing its transformation, and to help it claim greater status by demonstrating that it can play a greater international role than it currently is, and that it can do the same things great powers can.[3]

Self-strengthening is not a grand strategy that is likely to work for every rising power. Remaking the fabric of one's society is an undertaking that is politically costly and likely to generate opposition from traditionalists, as well as from those groups asked to shoulder the costs of reforms. For this reason, self-strengthening is most likely to succeed under authoritarian and centralized conditions, in which opposition can be ignored, silenced, or subjected to repression, and in which nationalist ideology or the charisma associated with the top leader can be used to justify sacrifices of money, values, and blood from the population. By contrast, a democratic or decentralized state with a weak government would find it hard to achieve the necessary domestic political consensus to enact or keep in being the reform program, because the need to maintain a domestic coalition strong enough to stay in power would require compromising the extent or duration of the reforms.

Another common trait of rising powers that adopt self-strengthening has to do with their unique context as states that own large latent resources, especially in terms of a commensurable population and territory, but which have failed to be exploited to their full potential because of the relative backwardness of the state in terms of the efficient organization and extraction capacity, hence social and political power. A state that lacks in latent capabilities may profit little from self-strengthening, because its problem is not one of efficient husbandry of resources, but lacking them in the first place.[4] Meanwhile, a state that is more advanced in terms of political and social power does not require a radical transformation in order to achieve competitiveness with better-off actors.

Opportunity strike and fait accompli

"Opportunity strike" refers to a grand strategy that seeks to exploit windows of opportunity produced by a favorable international environment in order to achieve gains at the expense of another state through the use of military force, whether through conquest, coerced inclusion in its sphere of influence, or the imposition of a client government. This grand strategy is characteristic of Frederick the Great's Prussia, Cavour's Italy (Sardinia), the US in the 1890s, and arguably Nazi Germany. Opportunity strikes are conducted against targets that are either temporarily or chronically vulnerable and which, therefore, are not likely to put up serious resistance.[5] Opportunity strikes are not bolts out of the blue, or surprise attacks aimed at defeating a powerful enemy before it is ready for a fight. Instead, they are efforts to achieve a determining advantage that levels the field against other great powers by seizing control over the resources of a weak target and adding them to one's own.[6]

There are two key elements in an opportunity strike.[7] The first is timing, as the rising power wants to ensure that one or several great powers will not step in to rescue the target. Thus, military advantage against the target alone is not sufficient for a rising power to conduct an opportunity strike. Instead, rising powers look out for intervals in which the states most likely to intervene are tied down somewhere else; in which the target is so weak as to invite a general onslaught, and hence, there is high probability of having it face several challengers at once; in which potential guarantor great powers write off or weaken their commitment to rescue the target; or in which the target is resented by most great powers, so they would not act to help it out.[8] Furthermore, timing is also important in the sense that such windows of opportunity may close quickly. Great powers may change their original passive stance when discovering the aggressive intentions of the rising power, which makes it all the more important for it to secure its gains before counter-intervention may occur. Accordingly, recognizing and exploiting windows of opportunity in a short timeframe are essential for the success of an opportunity strike.

The second element in an opportunity strike is decisiveness. On the one hand, opportunity strikes are successful only if they achieve decisive results, that is to say, only if they are fast, cheap, and complete. If the rising power becomes caught up in a long-term costly war of attrition, this increases the likelihood of outside counter-intervention, and makes it harder to seize and exploit the target's resources. On the other hand, the gains produced by the opportunity strike in terms of seizing control of the resources of the target have to be large enough so as to help solve the rising power's problem of lacking capabilities, or spheres of influence comparable to those of other great powers. This is achieved through the acquisition of territory of critical military value, especially enabling defense, or of vital natural, industrial, or demographic resources. As a rule, empires throughout history have not prospered due to the strength of their core, which was limited, but rather because of their ability to cumulate or substitute the resources of their conquests with their own, thus continually increasing the size of their population, military, and economies.[9]

The opportunity strike is, however, only half of the grand strategy, because the rising power needs as well to make other great powers accept its gains as legitimate. This not only entails that they should de facto put up with its control over the target by refusing to challenge it by force, but also that they should admit openly that the rising power has the legitimate right to exercise control there. As long as this is not the case, any gains can be reverted in the future, particularly when tables are turned against the rising power. This is why the rising power needs to combine military and diplomatic elements in making others accept the facts that have been created, or, in other words, put up with the fait accompli. Militarily, the rising power should be prepared to fight to conserve its gains, while diplomatically it should seek to gain allies so as to prevent being confronted with an overwhelming united front of opposition. This implies that the rising power must offer concessions to other powers in return for their recognition of its gains.

An important feature in getting a fait accompli to be accepted is that a state that has conducted an opportunity strike cannot resort to another one for some time in its aftermath, because other states will view it with suspicion. Conducting another opportunity strike too soon after the first will only confirm suspicions that the state is a serial predator and would lead to other states joining forces to contain or roll back its advance. That is why the rising power needs to deny any further ambitions and refrain from any aggressive moves for at least as long as its claim over the gains from its previous effort has not been fully accepted by the other great powers.

Opportunity strike and fait accompli is a grand strategy that is likely to be particularly attractive for rising powers that are either lacking geopolitical, demographic, and economic resources on par with other great powers, but have on the other hand powerful militaries, or for late arrivals in the great power club who cannot expand in an unclaimed area and, therefore, have to carve one out of weaker parties and of portions of the spheres of influence of other states. Opportunity strike and fait accompli tend to be favored therefore by rising powers that are weaker or disadvantaged, and who seek to level the playing field, not, as is the case with self-strengthening, through better management of their own resources, but through commandeering the resources of others.

However, opportunity strike and fait accompli can only work in a world where resources are cumulative, that is to say in which states can easily add the resources of other states to their own. This is likely to be increasingly hard in present circumstances, first due to nationalism, because the target state may mount a stubborn long-term resistance against the rising power even if odds are against it, and in doing so increasing the costs of control. Second, resource cumulation is hard in intervals in which conquest is problematic, because its costs are raised by defensive or deterrent technology such as castles, fortifications, the machine gun, or nuclear weapons, or by attrition military strategies, such as guerilla warfare.[10] Third, resource cumulation is also made harder by globalization, because wealth is tied to a global chain of production, in which raw materials, parts, technical knowledge, assembly factories, and sales are spread out around the world. This means that controlling any link in this chain does not result in greater resources, since to extract profit, all the links would need to be controlled. Even taking over a country would not guarantee acquiring its bank accounts, which would quickly dry out as funds are being digitally transferred abroad.[11] Finally, opportunity strike and fait accompli is perhaps the riskiest form of strategy for a rising power, as one may miscalculate the likely reactions of other great powers or may fail to reconcile them to the fait accompli, turning what promised to be a quick and decisive fix to the state's problems into a debacle. Warning examples are Nazi Germany in 1939, and Iraq's attempt to occupy Kuwait in 1990.

Tertius gaudens and divide and conquer

This grand strategy is enacted when a rising power, instead of waiting for the opportune moment to expand, takes measures itself to create propitious conditions for increasing its power and status. This is done by encouraging and exploiting the rifts existing between the great powers, so that the rising power ends up benefitting from their conflict and disunity. The historian Geoffrey Blainey referred to such a situation as the fighting water birds dilemma: two water birds fight over a fish, which in the meantime a fisherman snatches away.[12] The third actor in a relationship stands to profit from the conflict between the other two, hence the label tertius gaudens, meaning rejoicing or benefitting third. This grand strategy was especially evidenced by Bismarck's Prussia and Germany.

For the rising power, conflict serves several purposes simultaneously. First, it works as a wedge or an anti-balancing strategy, aimed at preventing the creation of a coalition of hostile great powers, or at breaking such a coalition. Because they view each other as bigger threats than each of them views the rising power, and, consequently, mistrust each other, great powers have difficulties ganging up against the new arrival. Second, conflict also has a deterrent role. Since it has to worry not only about the rising power, but also about other more formidable enemies, any great power, which otherwise would seek to block its rise, faces the risk of getting caught in a wider, more expensive, and more dangerous war. While the great powers hold each other in check as in the fighting water birds dilemma, the rising power can increase its power and status with impunity. Third, conflict also enables the rising power to gain helpful allies and partners. In order to defeat their main enemy, they are willing to provide it with military, economic, and technological assistance, as well as with recognition of its status claims.[13]

Tertius gaudens is pursued chiefly by means of alignment, which means taking sides and in so doing, giving rise to expectations both of support and of confrontation for certain parties. The logic here is that if states A and B are at odds, and C sides with A, it is expected to offer support to A and to confront B. But if more states and issues are involved, alignment becomes far more complex. C may side with A on some points, but with B on others, or if another state D is at odds with A, it may support either A or D. Therefore, what the rising power does is to repeat this step of taking sides in regards to other powers or/and other contentious issues in such a way that it ends up in the center of a network in which every conceivable action against it is likely to generate a counter-reaction by another great power.

At that point, every great power needs its support and none can afford to lose it, so it can ask its price in terms of an increase in power and status, or take action against a third party while its partners are busy confronting each other. Divide and conquer takes tertius gaudens one step further. Instead of merely joining one of the sides in a conflict already under the way, the rising power incites a great power to take an action that would make it fall out with another great power, thus increasing the value of the divider's support on that issue.[14]

Why is the rising power successful? One reason is that it relies on a combination of carrots and sticks, the former in the form of promises of support; the latter, in the form of threats of confrontation or of withdrawing support. Accordingly, the rising power can neutralize a would-be opponent by offering it assistance on an issue that it particularly cares about, removing it from an eventual counter-coalition; or it can bribe it by offering it side payments at the expense of a weaker third party; or it can threaten it with joining its chief enemy; or it can blackmail its current partner with the threat of abandoning it in its hour of need. This ability of the rising power to successfully play great powers against each other relies, in turn, on its flexibility.[15] Since the rising power is able to join either camp, it enjoys considerable bargaining leverage, since it can put up a convincing case that if it is not content with the terms it is getting, it can simply switch sides, acting as a pivot in the relation. The rising power also capitalizes from its relative weakness and lower status compared to either competitor. Being less formidable and ambitious than either side, it pays for them to accommodate its rise. Doing so allows the great power to recruit the rising power against a more formidable opponent. Moreover, with an on-going conflict against a rival in the same league on its hands, the great power cannot afford throwing resources away so as to block the rising power's ascension.[16]

Tertius gaudens and divide and conquer can succeed only under certain conditions. First, the great powers must both be at odds with one another and remain so. If they ever manage to settle their differences or strike an accord, the opportunity for the rising power to play them against each other vanishes. This was the case in the 1900s, when German decision-makers found out to their consternation that the seemingly intractable former rivals Britain on one side and Russia and France had concluded understandings, or ententes, with each other. The same scenario occurred at the end of the Cold War, when due to American–Soviet reconciliation, a host of Third World States in the Middle East, East Asia, Central America, and Africa, which had extracted concessions from the two sides during their competition, found that superpower support had suddenly dried up. Second, for the grand strategy to work, the great powers must fail to realize the rising power's real game. If they are alerted to the fact that it seeks to set them against each other, or, so to speak, to get them to pull its chestnuts out of the fire, they are more likely to resort to rapprochement and to cut their ties to it.[17] Third, the rising power should avoid compromising its flexibility. If its enmity for a particular great power becomes taken for granted, it can no longer convincingly threaten that it may switch sides. Consequently, a great power that shares the same enemy with the rising power has no longer a reason to pay for its support, since the rising power has no real choice but to stay in its camp. Imperial Germany in the 1900s is again a good example: its threat to Great Britain that, should it offer no concessions on colonies, it would have joined forces with France and Russia appeared as a bluff because, due to the rivalry between Berlin and Paris, there was no chance that it would have taken that step. As the British commented, there was no reason to pay Germany for something that Britain was going to enjoy anyway.[18]

Reassurance and biding your time

Reassurance and biding your time occurs when a rising power seeks to assuage the concerns of great powers about its ascension by adopting a conciliatory position which prevents, delays, or minimizes an adverse response.[19] This is best seen in the grand strategy of China under Deng Xiaoping and his successors Jiang Zemin and Hu Jintao. Scholars generally agree that the key components of assessing threats comprise power, especially military power, and intentions. Obviously, a rising power cannot pretend its power is not increasing, but it can manipulate how its intentions are being perceived by other great powers. Thus, it seeks to convey the image of a non-aggressive state, which plays by the rules, has no demands toward third parties, nourishes no ambitions of improving substantially its status, and can be trusted to cooperate with the great powers instead of challenging them. There are several ways to put out this message. First, the rising power will send a number of costly signals consisting not just of rhetoric, but also of actions that presumably an aggressor would not be able to take without sabotaging efforts at mounting an opportunity strike. The rising power would refrain from developing its armed forces or would cut them down unilaterally. It would consent to arms controls and verification measures and would adopt a defensive military doctrine. It would abstain from threatening force as a way to champion its interests. It would settle its territorial disputes with its neighbors, even if it means abandoning past claims. Assuming that there is a way to differentiate between offensive and defensive weapons, the former tending to be more mobile (tanks, bombers, aircraft carriers), while the latter are more static (fortified positions, fighter planes, destroyers), the rising power will avoid investing in weapons considered offensive. Finally, although this step was not fully taken by China, it could change its ideology, form of government, revolutionary practices, or treatment of minorities and dissidents to make it more compatible with that of the great powers, and, hence, more tolerable.[20] Second, the rising power would seek to join regional and systemic institutions and would make a point in scrupulously respecting their rules, even when they leave it at a disadvantage. Since the objective of these institutions is defending the existing order, participation and support for their norms are seen as a signal of accepting the legitimacy of the overall international arrangements.[21] Third, the rising power would pursue a "charm offensive," aimed at showcasing its cultural achievements, as well as at encouraging cultural and educational exchanges. The goal is to create sympathy and goodwill among public audiences, especially academic ones and in the public opinion, which would complicate the mission of a decision-maker advocating taking harsh measures against its rise.[22] Fourth, it would make other great powers reap benefits from its rise by opening its markets for international trade and investment, and making these profits conditional on others putting up with its ascension.[23] Fifth, the rising power would pass up opportunities to claim additional status. As such, it would downplay its success, overemphasize its weakness, refuse additional responsibilities, and decline participation in a leadership role.

Reassurance and biding your time is based on the rising power's conviction that time is on its side, so, for as long as it successfully avoids conflict, its capabilities will keep rising to the point where the opposition of other great powers would no longer matter. Concessions can therefore be liberally dispensed with the knowledge that once it becomes the more powerful party, the rising party can denounce them or recover them by extracting in turn concessions from the great power, now in the position of the weaker party.[24] In effect, the rising power is playing for time, accumulating resources under the very nose of the great powers. Why is it that the great powers do not realize what is going on and take preventive action against the rising power? The reason is that due to the rising power's concerted efforts, such action is seen, at the same time, as unnecessary, because the target shows itself as moderate; premature, because it presents itself as too weak to warrant it in the here and now; and self-defeating, because it would hurt one's own economy by killing the goose that laid the golden eggs.

Reassurance and time-biding work best for a rising power that would take a long time to complete its rise, that is to say, for a state that has to start rising from genuinely humble beginnings. The rising power's message that it does not represent a threat cannot help but be convincing, because it has a base in reality. Being genuinely weak, it would take a long time to prove a challenge. The same message would not be as believable coming from a particularly strong great power vying for the number one spot. For instance, it did not matter that Imperial Germany's decision-makers before World War I proclaimed their intention of navigating the danger zone of ascension to world power carefully, like a butterfly emerging gradually from a caterpillar's chrysalis. Germany was simply too big and ambitious to pretend credibly that it was weak and non-aggressive.[25] Reassurance also requires a congenial international context, since it leaves the rising power vulnerable from a military standpoint to depredation by the great powers. For reassurance to work, the international situation must be such as to dissuade the great powers or other rising powers from taking advantage. This is more likely in periods of uncontested international order underlined by a dominant state or a concert of powers than in intervals of transition when the order breaks down and there is a free-for-all among the great powers to secure benefits. Furthermore, reassurance and time-biding, like self-strengthening, are more effective in a non-democratic setting. Concessions to other states and great powers, especially in terms of territory, military power, and ideology, are likely to be resented by domestic audiences as defeats and national humiliations, which would put a democratic government at risk of being unseated in the next elections. The public is required to remain quiescent and patient in return for the fulfillment of state ambitions in the distant future. This is easier to accomplish if political dissent is restricted.

Reassurance and time-biding also suffer from several potential shortcomings. Time-biding succeeds to the degree that time is really on the rising power's side. But in the long run, friction or the unforeseeable is more likely to occur, making the rising power lose its bet on the future. The economy may stagnate, technology may be made obsolete by innovation spearheaded by competitors, population may decline, a general war may break out, or the government may be weakened or even toppled by domestic unrest. In 1956, the Secretary General of the Communist Party of the Soviet Union Nikita Khrushchev proclaimed that history was on the Soviet Union's side and that, inevitably, the Soviet Union was going to bury the West.[26] Less than four decades later, the Soviet Union no longer existed. Furthermore, while the rising power may offer concessions to signal its acceptance of the international order, it is never able to offer the most convincing concessions of all in the form of completely curtailing its rise. Accordingly, no rising power would be able to give up definitively increasing its economic resources, refrain indefinitely from translating them into military power, and then reject for all time claiming additional status. To demand such concessions from the rising power is bound to produce its resistance, which would leave reassurance in shambles.[27] Finally, reassurance and time-biding can never be more than a temporary strategy. While it is tempting for a rising power to avoid opposition to its rise by continually keeping its head down, this is not possible in the long run if it wants to satisfy the criteria of behavior and recognition as a great power. That is to say that eventually a rising power has to demonstrate its credentials as a great power by acquiring new spheres of influence, prevailing in wars and confrontation, claiming a wider share of system management, and undertaking a larger number of international commitments. But such actions are fundamentally at loggerheads with staying out of trouble.[28]

Self-strengthening: the rise of Russia

Russia in 1672, the year of Peter the Great's birth, was an unlikely candidate for ever becoming a great power. While it was already the largest state in Europe in terms of size and the second largest after France in terms of population, at an estimated 10–15 million, Russia's natural resources were unexploited, its population was spread thin across its vast territory, and its large human capital was underdeveloped. The tsar was not even aware of the full extent and layout of his realm, which had not been properly mapped. Russia had a single major city Moscow, with 150,000 inhabitants. The second largest was only a tenth of the size. There was no industry to speak of, and foreign traders surpassed in number Russian merchants. If Russia could field on occasion a large army of up to 100,000 men, its equipment, composition, organization, and tactics were antiquated by comparison to the great powers of the day. Russia had tried and failed, once before, to become a great power during the reign of Ivan the Terrible, but ended up defeated and cut off from the Baltic Sea. In 1672, Russia was checked in the south by the Khanate of Crimea and its suzerain the Ottoman Empire, while in the west by Sweden, a military powerhouse believed to possess the qualitatively best army in the world, controlling not only present-day Swedish territory, but also Finland, the Baltics, and part of Germany. Consequently, Russia was perceived as a state of marginal importance. In the 1648 Treaties of Westphalia, it was briefly mentioned as adhering to Sweden.[29] How could a poor, inefficient, technologically and organizationally backward Russia transform itself in the space of half a century into a formidable great power?

The answer has to do with the self-strengthening grand strategy adopted by Peter the Great.[30] The tsar had a very unusual educational background. A rebellion of an elite musketeer corps (the streltsy) in 1682 came very close to unseating him. Peter had to watch how his chief councilor was hacked to death with halberds. In the aftermath of these events, Peter's elder sister Sophia assumed regency. This led to Peter undergoing a self-imposed exile from Moscow proper to the nearby village of Preobranzheskoe. There, he grew up by learning from the foreign merchants, craftsmen, and soldiers from the foreign suburb of Moscow the Dutch language, as well as a multitude of trades. He boasted he had mastered 14 crafts, among which were gunnery and shipbuilding. Peter already had a strong interest in military affairs, and developed a lifelong ambition in ensuring that Russia developed an army and navy on par with those of the Western powers. Even before he started reigning as a single ruler, Peter had organized his household troops in so-called toy regiments, armed, dressed, and organized on Western principles with which he staged mock maneuvers and battles and engaged in experiments in shipbuilding.[31] Then, in 1697, Peter took a completely unprecedented step for a Russian tsar, by leaving incognito at the head of a large retinue in order to visit firsthand the countries of the West. While the so-called Great Embassy's ostensible goals were to recruit shipwrights, officers, and navigators for Russia's budding naval program, as well as to convince the West to continue to pursue a common offensive against the Ottoman Empire, its major consequence was a change in Russia's grand strategy.

Peter discovered, by working as a common ship carpenter in the shipyards of Amsterdam and London, that it was impossible to create a Russian military and navy that could rival the best the West had to offer just by importing wholesale technology and expertise. The Western powers' military and naval superiority rested on deeper structural, economic, legal, administrative, educational, and cultural foundations. Peter took this so much to heart that his own seal carried an inscription that read: "I am a student and I seek teachers." That meant that Russia had to develop a similar array of institutions and practices in order to be able to eventually compete at the same level.[32] Thus, Peter sought to transform Russia into a modern European state by the standards of the day.

This was no mean feat because it required building up both the idea and the machinery of the modern state from scratch. Prior to Peter, Russia had been lacking the idea of "the State, as a power taking upon itself the administration of all aspects of human activity for the aims of the common good." Accordingly, Peter's key motive in his many decrees and proclamations was the conviction that all people, himself included, lived to serve and advance the interests of the Russian state or "the common good." This was not just empty rhetoric. Peter took pains in replacing in official documents the formula "the interest of His Majesty the Tsar" with "state interest," and went so far as to eventually subject his own son Alexei, who was favoring rolling back the reforms and returning to tradition, to arrest, torture, and, in all likelihood, summary execution.[33]

Self-strengthening was carried out through a crash program of reforms imposed from the top down. The gist of these measures was to create the tools for the government to mobilize and organize efficiently the demographic resources of Russia, by using Western technology and organizational methods. While employing foreign advisors and experts in the initial reform phase, Peter preferred relying on Russians that had mastered Western knowledge and methods. Military reform represented a central concern, swallowing up to 70% of government expenditure. Peter started by conducting the first census of the Russian population. He then reformed the army recruitment system based on the model of Sweden, with every 100 households being forced to provide one man for the army. By 1725, Russia could deploy an army of 200,000 uniformed troops armed with Russian-produced guns and cannons and commanded by trained Russian officers. Therefore, the reason for Peter's Russia's new-found military strength had to do with its ability to mobilize large number of soldiers and to do so continually. It is estimated that no fewer than 331,000 men were mobilized from 1699 to 1714, at a rate of 22,000 per year.[34] Since this new modern army needed to be equipped, clothed, and staffed, Peter spent a further 10–15% of his budget to create a state-owned network of textile and metal-working factories, opened the largest iron and copper mines in Europe in the Urals, built shipyards with the Admiralty in St. Petersburg being one of the largest enterprises in Europe employing 10,000 men, and created compulsory officer schools for the sons of the gentry. This was accompanied by a large-scale building program of canals, harbors, roads, and cities, the prime example being Russia's new capital at St. Petersburg. In addition, Peter pursued administrative reforms of the central and local government, established new military and civilian hierarchies, changed the calendar to correspond to the Western system, opened Russia's first technical schools and its Academy of Sciences, sent exploration missions to Siberia, became a patron of the fine arts, and commissioned the writing of the first history of Russia.[35]

The roots of Peter's success were three. First was a vastly improved system of taxation, again based on the census, which allowed Russia to self-finance much of its development. As Peter put it, "money was the artery of war." In addition to regular taxes per person or soul tax, Russians were paying army taxes imposed on particular social classes and categories, such as warship money or dragoon money. But Peter eventually subjected to taxation almost every activity and luxury, levying for instance taxes on items such as fish nets, beehives, bath-houses, glass, doors, and oak coffins. Even wearing beards, which was frowned upon, since Peter sought to enforce Western clothing and a clean-shaven hairstyle, was taxable. The result was that states' revenues went up five times between 1680 and 1724, from 1.5 to 8.7 million rubles.[36] Second, such radical transformation would have been inconceivable without the ability to conscript at will tens of thousands of peasants for service in the army, or for labor without wages in factories, mines, and constructions. It is estimated that no fewer than 40,000 peasants were levied per year solely for the purpose of building St. Petersburg on drained marsh terrain.[37] Third, it was no secret that in order to promote reforms of such magnitude, dissent had to be quelled. This was done on several occasions, most notably in a crackdown on the streltsy for an aborted revolt that cut short Peter's Great Embassy in 1698, and in a punishing expedition against the Zaporizhian Cossacks who had sided with Sweden. However, in a larger sense, all of Peter's reforms were based on coercion, a subject about which he was quite open. As he commented,

> they call me a savage ruler and a tyrant ... but who says this? People who do not know the circumstances in which I found myself a few years ago do not know that many of my subjects placed the most foul hindrances to carrying out my best plans for the benefit of the fatherland, and, therefore, it was essential for me to treat them with great severity.[38]

The second part of Peter's self-strengthening program involved using Russia's improved capabilities to secure a proportional improvement in its status by securing, at the same time, recognition as a great power. Originally, the tsar had contemplated expansion against the Ottoman Empire, but after the Great Embassy, it became clear to him that aggrandizement in the south could not contribute to the objectives of drawing closer to the West, economically, politically, and culturally, as well as of producing an impression of Russian strength on the Western European powers.[39] In order to do so, it was necessary to renew efforts in the Baltics against Sweden, and primarily to regain access to the Baltic Sea coastline. As Peter put it, Sweden "had not only robbed us of our necessary harbor, but in order to deprive us of the desire to see, has put a heavy curtain in front of our mental eyes and cut off our connections with the whole world."[40] In other words, for Russia to become an important actor in Central and Western Europe, it had to make Sweden give up the Baltics. Essentially, this meant that to become a great power, Russia had to be able to defeat Sweden, an acknowledged great power, in war.

It was at this point that Peter resorted to an almost fatal miscalculation. As he later admitted, he and his advisors had not understood the opposing forces and Russia's own situation, and began the war against Sweden "like blind men."[41] Effectively, this was a botched opportunity strike. On the surface, the war began in 1700 appeared a cakewalk. Russia could count on the support of Denmark and Poland (whose elected king also occupied the throne of the German state of Saxony), both of whom also wanted to increase their territories at the expense of Sweden. Sweden seemed particularly ripe for the taking, with both internal turmoil, and the accession to the throne in 1697 of a boy of 14, Charles XII. The perfect plan went awry, however, in practice. Charles XII proved to be one of the foremost military commanders of the age, and defeated piece-meal the attackers, beating back the forces of Poland-Saxony, and then knocking Denmark out of the war. Left alone to face a revitalized Sweden, Peter soon discovered that his reformed army was still not up to a contest against a modern great power army.

At the stronghold of Narva, Charles XII at the head of an army numbering only 9,000 routed a Russian army of 40,000, in a battle fought in a snowstorm, which made it impossible for the Russians to tell the numbers of the Swedish attackers. Russia suffered a stinging defeat, losing in the process 10,000 dead and thousands more being captured, not to mention its entire artillery.

Russia ultimately prevailing in its contest with Sweden, later referred to as the Great Northern War, had two reasons: the even greater mistakes of its enemy, and its own superior resources, now harnessed through self-strengthening. On the one hand, Charles XII erred by discounting Russia's military ability after Narva. Instead of concluding peace with Saxony or following up its victory with a fast offensive against a stunned Russia, the Swedish king sought to exact revenge upon the aggressors. Accordingly, he invaded Poland, and got bogged down for years in trying to impose his own candidate upon the Polish throne. This in turn allowed Russia time to recover and to continue and enlarge the self-strengthening program. Russia took again the offensive and was able to seize from Sweden, Ingria, the area around present-day St. Petersburg, and Livonia, a territory comprising parts of present-day Latvia and Estonia. There, Peter founded his capital in 1703. By now, Russia had a position of strength from which it could intervene to fight Sweden in Poland. Eventually, even Narva fell to Russian troops.[42] The second and more important reason was that Sweden was fighting the war with limited resources in terms of soldiers, which, once spent, could not be replaced, while Russia had troops to spare, and could continually get more under arms through Peter's new recruitment system. This meant that Sweden could not afford any wrong steps or risk catastrophe, while Russia could more easily come back from defeats by capitalizing on its superior manpower. Therefore, it was Peter's ability to use self-strengthening as a way to translate Russia's previously untapped advantage in population into military and economic strength that proved decisive.[43]

When Sweden invaded in 1707, its troops were outnumbered from the get-go.[44] Instead of standing and fighting a decisive battle as Charles XII expected, Peter resorted to highly successful scorched earth military strategy, setting fire to provisions, fodder, and crops in the field, as well as to buildings, bridges, and forests. If anyone was to provide the Swedes with food the person was to be hanged, and the village from which the food came was to be burnt. The Swedish army had to transport its dwindling supplies through miles of hostile territory and the harsh Russian winter, because no food or fodder was available on site. The result was that in June 1709, an army of 22,000 weakened and famished Swedes, all that remained of Charles' initial invasion force, met in battle a Russian army twice its size at Poltava. Russia's victory was total: nearly 7,000 Swedes were killed in battle, the survivors surrendered in the aftermath, and Charles XII fled with a small contingent for the protection of the Ottoman Empire. From this point on, Sweden was reduced to fighting a desperate last-ditch battle against a revived coalition of Russia, Prussia-Brandenburg, and Saxony.[45]

As the single most important military force left in the war, Russia could take its pick of the spoils. It was no longer a question of keeping only Ingria, the region around St. Petersburg, which Russia had been promised initially by Denmark and Poland. It took 12 more years of stubborn fighting, with Russian troops eventually moving into Sweden and Germany, but at the end of the war through the Treaty of Nystad (1721) Russia ended up in control of Livonia, Estonia, and parts of Finland as well as the paramount power in Poland, which became a de facto Russian client-state, mentioned in the treaty only as Russia's ally. What mattered was not the extent of these territorial gains, but their location on the map, which transformed Russia from a state confined to the periphery into a major player in Germany, Central, and Eastern Europe. For Russia there was no possible going back to the days before Peter, since it was now a constant participant in the affairs of the West. In effect, ever since Nystad and up to the present, with only fleeting exceptions, Russia has conferred priority on its interests in Europe over those in any other region.[46]

Russia's conclusive victory—it had managed to single-handedly end Swedish invincibility—coupled with its extensive capabilities, also meant a significant change in its status, from a marginal actor to a great power. If before the war, no permanent diplomatic missions existed in Russia, in its aftermath St. Petersburg counted no fewer than 21. The great powers now saw Russia as an actor to be reckoned with: while France sought its mediation in the war of Spanish Succession, Great Britain was uneasy about the growth of Russian influence in the Baltics and concluded an alliance with Sweden to resist it. Furthermore, now Russia saw itself the equal of the great powers, best seen in Peter's assuming in the wake of the war the title of Emperor (Imperator), which previously had been reserved only to designate the Habsburg ruler of Austria.[47]

Opportunity strike and fait accompli: the rise of Prussia

When thinking of Prussia in the first part of the eighteenth century, imagine a state formed of three parts (the duchy of Cleves and the County of Mark in the west; the electorate of Brandenburg in the center; and East Prussia in the east), each part noncontiguous to the other, and separated by other German states and Poland, effectively making the ruler of Prussia the "King of the border strips." Furthermore, Prussia in 1740 had a minuscule population for a great power, 2.2 million, compared with France's and Russia's, which each exceeded 20 million. Even Britain and Austria dwarfed Prussia's population by a factor of four. To cap it all, Prussia was also lacking in economic resources, with such poor soil that it was dubbed "the sandbox of the Holy Roman Empire," and with little industry to speak of. These adverse conditions fitted Prussia to be a second- and possibly third-rank power, more akin to the other German electorates Saxony, Bavaria, and Hanover than to the members of the great power club. In fact, its king, Frederick II (Frederick the Great), often joked that the state's coat of arms should have shown a monkey instead of a black eagle, since Prussia could only ape the great powers without genuinely being one of them.[48] However, by the 1760s, few would have questioned that Prussia belonged in the great power club.

What accounts for Prussia's miraculous success? Part of the answer had to do with the fact that at the time of Frederick's death, despite having the thirteenth largest population in Europe, Prussia was fielding the continent's third largest military force. By contrast to the great powers that had only a small proportion of the population around 1–2% serving in the army, about 4% of Prussians were enrolled as soldiers, thus enabling the kingdom to punch above its weight with a force of 80,000 men. As its contemporaries observed, it was not a country that had an army, but rather an army which had a country.[49] Another advantage was the quality of Prussia's soldiers, based on precision drilling and maneuvering as well as on iron discipline. As the saying went, Prussian soldiers should have feared their officers above the enemy. The result was that "a Prussian battalion became a walking battery whose speed in reloading tripled its firepower and so gave the Prussians an advantage of three to one" on the enemy.[50]

However, this increase in soldiers' numbers and quality would have counted for little had it not been for the change Frederick II brought to Prussia's grand strategy. He made Prussia into "a first class-state based on a material base more appropriate to a country of the second or even the third order." The traditional grand strategy of Prussia was to keep good relations with its neighbors, and in particular with Austria, the strongest German state. Over time, Prussia's support for its great power patron had come to be taken for granted, which is why Prussia under Frederick's father, King Frederick William I, was seeking to enlarge its army. Knowing full well how most German states were so cash-strapped that they had to abdicate having any foreign agenda of their own, and even resorted to renting soldiers for the side that paid best, the king was attempting to secure a measure of autonomy by gaining a better footing for bargaining with Austria. Consequently, he urged his son to continue the same line by maintaining peace and good relations with Vienna, while continuing to increase the size of the military.[51]

However, these limited objectives were not an option for Frederick, because they still left the fundamental problem of Prussia's scattered possessions unaddressed. As he put it in 1731, Prussia "must seek major goals and leave minor ones aside." Without further territorial acquisitions linking its three disparate fragments and providing additional resources, Prussia would always remain vulnerable to foreign pressure and teeter on the brink of disintegration. The remedy required was therefore not more of the same moderation, but an aggressive strategy of expansion at the expense of the other German states.[52] Essentially, in order to escape the problems inherent to the condition of being a second-rate power, Prussia had to become a great power.

The opportunity for expansion presented itself a mere five months after Frederick ascended the throne. In October 1740, the Austrian ruler and Holy Roman Emperor Charles VI died. This provoked a dynastic crisis in Austria, because Charles had not left any male heirs. While Charles had taken precautions so that his daughter Maria Theresa could inherit the hereditary Habsburg lands, as a woman she could never succeed to the seat of emperor, which had been for 300 uninterrupted years always occupied by a Habsburg. Furthermore, Austria was bankrupt and its military was in disarray: its latest war against the Turks had ended up in an embarrassing draw, and even though on paper its strength stood at around 180,000 men, in practice it was estimated that its army could count on only half as many.

Frederick concluded that Austria was in poor condition to put up a fight and was going to become increasingly vulnerable, as more states were likely to put forward rival candidates for the imperial crown and to wrest away territorial gains.[53] By remaining passive in the upcoming scramble for power and status, Prussia would have failed to gain a share for itself. As he put it to his dissenting advisors, this would have been a great error:

> I am giving you a problem to solve. When one is in a favorable situation, should one make use of it, or not? I am ready with my troops and everything else; if I don't take advantage of it, I shall have in my hands an asset I don't understand how to use; if I take advantage of it, it will be said that I know how to make the superiority I enjoy over my neighbors work for me.

Later on, in his Political Testament of 1752, he expressed the view that a disinterested power sharing a neighborhood with ambitious powers would perish for certain. Therefore, he thought that it was vital to take advantage of Austria's weakness and move before any other states had the chance to do so.[54]

Frederick calculated that with the other great powers either at war with each other (France and Spain against Britain) or involved in domestic problems (Russia), there would be no foreign help available to Austria. Consequently, he decided to seize the opportunity presented to him in a matter of days. On October 26, the news that Charles had died had broken. By November 7, the order to ready the army for war was given. On December 16, Prussia invaded Silesia, a hereditary province of Austria. It mattered little that Prussia had no historical claims on Silesia. Frederick was quick to praise his chancellor, who hastily conjured a legal justification, as an "excellent charlatan." What did matter was that by acquiring Silesia, Frederick increased Prussia's population by a third; doubled its revenues by gaining a province rich in agriculture and with a prominent linen industry, which had provided the largest tax income for the Habsburgs out of all their provinces; and gained a valuable bridgehead for staging potential offensives against Austria. Not in the least, by wresting a province from Austria, an established great power, Prussia and by implication Frederick, gained status: as he emphasized, the invasion constituted a "rendezvous with glory" and his key motivation, as the French Ambassador to Berlin remarked, was the "love of grandeur, glory, and especially anything that can enhance his reputation among foreign nations." As Frederick predicted, Silesia was lightly defended, and the Prussian attack was followed by an onslaught of claims against Austria by Bavaria, Saxony, Savoy, and Spain, all of which were backed by France. [55]

Frederick had expertly seized the opportunity, but what was required now was to get the other great powers and in particular Austria to accept the fait accompli of the incorporation of Silesia into Prussia, an objective that eventually took more than 20 years and two costly wars to accomplish. For Frederick, the chance of further acquisitions in his own lifetime was slim, because after Prussia's surprise attack the other states would have been distrustful of its future intentions. As he wrote, taking Silesia was a onetime deal comparable to book stories, whose original had been a great success, but whose retellings by imitators were bound to fail. This did not mean that Prussia should have sworn off any future annexations, only that it should only have attempted them by dissimulating and hiding such intentions, and by waiting patiently for the right set of circumstances. Prussia should have preferred less expensive negotiations to the costly use of arms, if need be gobbling up new provinces like eating an artichoke, leaf by leaf. [56]

In the meantime, Frederick's chief concern was to preserve Prussia's hard-earned gains. In this, he had no illusions about his means: Prussia was still outnumbered by a combination of potential opponents. Nonetheless, Frederick could stay afloat by making good use of his assets, primarily his army now enlarged to 150,000 men, by pursuing alliances against Austria, and by conducting war with a careful eye on his war chest. The conduct of the War of the Austrian Succession on the part of Prussia was a masterful lesson of husbanding resources in accordance with necessities. Not only did Frederick employ mainly Silesian resources and men to pursue the fighting, but he also sought to explore a settlement with the embattled Maria Theresa that would have guaranteed him his acquisition. By 1742, after suffering defeat to the Prussian army at Mollwitz, and faced with enemies on multiple fronts, Austria had to offer concessions, recognizing Prussian rights to part of the province. Instantly, Prussia, which now had secured an alliance with Austria's archenemy France, pulled out of the war. But the very moment Austria's fortunes improved to the point of winning the war in 1744, which bode ill for Prussia's control of Silesia, Frederick rejoined the fray. By the end of 1745, Prussia had inflicted three additional defeats on Austria, forcing it to offer through the Peace of Dresden acceptance of the Prussian rights to the whole of Silesia, and again prompting Frederick to withdraw from a war that was to rage on for a further three years. By now Prussia had shown itself not merely equal but superior to the great powers on the battlefield as the owner of the most efficient army in the system; and its resources had expanded considerably due to the annexation of Silesia. Consequently, for all intents and purposes, Prussia had become a great power, a reality acknowledged when the two greatest powers of the day, France and Great Britain, competed to offer an international guarantee of the Prussian treaty with Austria.[57]

However, as Frederick realized, this was only to be the opening act in ensuring the fait accompli. Once she managed to overcome her difficulties by having her husband elected as emperor in the aftermath of the death of the main pretender, Maria Theresa's chief goal was to recover Silesia. In this project, she could also count on Russia, who regarded an emergent Prussia as a dangerous competitor in its own quest to influence Poland, and who consequently sought to cut Frederick's kingdom down to size.

It is here that Frederick miscalculated the diplomatic equation on which securing the fait accompli depended. Prussia needed allies to help secure its position. Frederick assessed correctly that France and Great Britain were set for another major war, which was probably going to sweep Prussia along as France's ally. Yet, he had misgivings about the extent of support he could expect from France. Thus, when Great Britain offered a subsidy to Russia in return for a security guarantee to the electorate of Hanover, the hereditary possession of the British king on the continent, Frederick feared the constitution of an alliance between Britain, Russia, and Austria directed at him, which he would have to face alone. Hence, he sought to nip in the bud the prospective combination by signing a similar treaty with Britain.[58]

However, for the court of Louis XV, Frederick's latest maneuver was a stab in the back by a king who had shown himself to be an unreliable ally in the last war, having twice left France in the lurch. Now, Frederick was openly consorting with the sworn enemy Britain, even going so far as rendering impossible a French offensive against Hanover, the one British possession that was vulnerable to French armies. The fact that Frederick was scathingly critical of Louis' mistress Madame de Pompadour, who was one of the main voices influencing France's foreign policy, only reinforced French enmity. Simultaneously, a rapprochement was taking place between France and Austria, who now considered Berlin, and not Paris, to be its chief opponent. The new Austrian chancellor Kaunitz advised, consequently, forming an alliance between the three biggest continental powers, France, Russia, and Austria to take the fight to Frederick. Thus, the Prussian king ended up causing the very situation he had wanted to avoid in the first place: diplomatic isolation and invasion by a numerically superior coalition. Prussia, a state of about 5 million people, was about to take on a combination of powers with a total population of around 90 million.[59]

Yet, with the diplomatic pillar of the fait accompli now gone, Frederick had to rely exclusively on the military one. He could not count on a decisive victory, but on prolonging the fighting long enough to force the enemy to settle. As the French foreign minister commented, "the king of Prussia does not want to lose, because if he doesn't lose, he will have won everything." Frederick's one advantage was that being warned by intelligence that an attack against Prussia had become inevitable, he moved to preempt it by yet another opportunity strike, this time seizing the German state of Saxony, which was bound to join the coalition. While on the surface the move backfired, because now the rest of the Holy Roman Empire, as well as Sweden, threw in their lot with the coalition, the occupation of Saxony made strategic sense. As Frederick put it, not even the most stupid prince would sit back and wait for his enemies to carry out their plans to destroy him. It was not for nothing that Frederick had marked Saxony as Prussia's most desirable next acquisition in his Political Testament. Prussia added the commensurable resources of the occupied electorate to its own and, effectively, waged war at its expense. Saxony ended up supporting more than a third of the Prussian war costs. The Saxon resources enabled a pragmatic war plan, in which Frederick abandoned indefensible Prussian territory in the east and the west, and, instead, focused on the defense of the center of the kingdom, where his own lines of communication would be shorter than the enemy's. By a series of fast marches and maneuvers from one front to the other, he could a) face each opponent separately with the same men so that he would never be decisively outnumbered, b) defeat them piece-meal, and c) prevent them from ever joining forces. He compared this military strategy to a trapeze artist on a high wire inching toward safety while ever faced with the prospect of disaster.[60]

Against all odds, Prussia survived. Even so, it was a close thing, and due as much to Frederick's superior planning as to the errors of his opponents and to the unforeseen. While Frederick scored a series of impressive victories, the most famous of which, Rossbach over France and Leuthen over Austria, occurred only one month apart, at the end of the day, the coalition could replace its losses much easier than Frederick could. In some battles, the Prussian army suffered casualties of nearly 30% of its total fighting force Even though, after Rossbach, the French preferred to concentrate their efforts on the colonial war against Britain, Prussia still had to face Austria and especially the enormous Russian armies. There were times, especially after a nearly disastrous defeat at Kunersdorf, in 1759, at the hands of the Russians, when Frederick himself thought all hope was lost to the point where he contemplated suicide. Out of 16 battles, he managed to win only half.[61] Chance however finally smiled on Prussia in 1762, when the Russian Empress Elizabeth I died. She had been a sworn enemy of Frederick, but to her heir Peter III, the Prussian king was a hero and subject of adulation, to the point where he was regularly wearing the Prussian uniform. Russia ceased hostilities against Prussia, and, incredibly, the perspective was even open for a Prussian-Russian alliance against Austria. It was not to be, as Peter was overthrown, and then assassinated in a plot orchestrated by his German wife who became the Empress Catherine II (the future Catherine the Great). The new Russian ruler subsequently pulled out from the war, meaning that effectively Austria was the only power left to confront Prussia. By 1763, even Maria Theresa understood that Austria could not defeat Prussia alone and by the Peace of Hubertusburg, she settled on the restoration of the status quo before the war started.

While this meant that Frederick had to restore Saxony, in the process, he had secured the possession of Silesia. He could also boast having fought the rest of the great powers to a standstill, since the support from his nominal ally Great Britain had been limited.[62] Thus, Prussia's performance helps to account for von Ranke's requirement, mentioned in Chapter 2, that for a state to be a great power, it should be able to maintain itself against all the others, even when united.

Divide and conquer: the rise of Germany

By 1862, Prussia was in better shape than its eighteenth-century counterpart. Its geopolitical, demographic, and economic situation had ameliorated substantially through the incorporation, as a consequence of the Napoleonic Wars (1803–1815), of both a substantial portion of Poland, as well as the Rhineland and Westphalia. As a result, its status as a great power had been uncontested since the Congress of Vienna (1814–1815). However, Prussia remained firmly at the bottom power in the club. The 1860 *Times* summed up the common view that Prussia was

> always leaning on somebody, always getting somebody to help her, never willing to help herself. … She has a large army, but notoriously one in no

condition for fighting. … No one counts her as a friend; no one dreads her as an enemy. How she became a great power, history tells us; why she remains so nobody can tell.[63]

Yet by 1871, Prussia had unified Germany and reached "a degree of material primacy in Europe which no great power had achieved in Europe since 1815."[64] What is all the more remarkable, it achieved this objective without ever triggering a coalition of the other great powers aimed at slapping it down, a development that according to Henry Kissinger was to be expected "according to all traditions of Realpolitik."[65] The years after 1815 were dominated by the so-called Concert of Europe, of which more will be said in the next chapter, composed of the five great powers (Great Britain, Austria, France, Russia, and Prussia). The members of the Concert showed considerable vigilance whenever another member risked gaining a decisive advantage in power over the other members, and, more than once, acted to block such moves, including by the threat of joint intervention. How then did Prussia manage to become overnight a frontrunner?

This commensurable success is largely attributable to the grand strategy of the Prussian chancellor Otto Von Bismarck. For Bismarck, the main objective was securing Prussian supremacy over the other German states. This goal was likely to meet with resistance from Austria, which meant that, sooner or later, a reckoning between Berlin and Vienna was unavoidable. As Bismarck put it in 1855, Prussia and Austria were "breathing each other's breath," meaning that one of them must "yield or be forced to yield by the other." In 1862, on the eve of being made chancellor, Bismarck confided to Benjamin Disraeli, the future British prime minister, what his foreign program consisted of: "my first care will be to reorganize the army. … As soon as the army shall have been brought into such a condition to inspire respect, I shall seize the first best pretext to declare war against Austria … subdue the minor states and give national unity to Germany under Prussian leadership."[66]

The way to achieve this result was to exploit and heighten the existing conflicts among the other great powers in such a manner that Austria would end up isolated. Should any great power have contemplated coming to its aid, it would have run into the opposition of another great power. As Bismarck put it, the ideal was a situation "in which all the powers … had need of us, and would thus be deterred as far as possible from coalitions against us by their relations with each other."[67] This was effectively a divide and conquer strategy.[68] By embroiling and maintaining at odds the other contenders, he ensured that they could not ever join forces against Prussia, and later on, against Germany.[69] As he put it, "all politics reduce themselves to this formula: to try to be one of three, as long as the world is governed by an unstable equilibrium of five Powers." Therefore, the fulcrum of his grand strategy was to avoid being isolated, and at the same time isolate opponents by perpetually leaving them in the minority. Actually, Bismarck was not aiming for just ensuring a perpetual three-versus-two safe majority for Prussia, but for a far more comfortable four-against-one scenario, in which Prussia/Germany's main enemy was always the one to find itself alone. The other three powers would have ended up due to their enmity toward each other as allies, partners, or neutrals favoring Prussia, depriving the target state of any support.[70]

The occasion for Prussia to advance its influence over the rest of the German states came in 1863 with the death of the Danish king. His successor, Christian X sought to bolster his domestic support by ratifying a new constitution which would have included the duchies of Schleswig and Holstein into Denmark proper. Legally speaking, that was a problematic endeavor. Holstein, unlike Schleswig, had a German, not Danish, majority, and was part of the German Confederation, which comprised all the German states. Moreover, Denmark had agreed through the Treaty of London of 1852, guaranteed by all the great powers, that it would not seek to alter the status of the duchies, which was ostensibly what was taking place. Hence, profiting from an incensed German public opinion, which rallied to the cause of its conationals in the duchies, Prussia sought military intervention.

Bismarck sought to undertake the action in partnership with Austria. This was very shrewd. On the one hand, Austria had nothing to gain territorially by seizing the duchies, which were not contiguous to its possessions. On the other hand, Austria could not excuse itself from taking part in the intervention against Denmark without jeopardizing its prospects of leadership of the German states in favor of Prussia. By this move, Prussia was gaining Austrian support against the other great powers—as Bismarck put it, "I shall not take a step without her [Austria] sharing our risks and perils." The joint intervention by the two German powers created, in turn, a conundrum for the rest of the powers in the club. Russia was still smarting from its defeat to Great Britain and France a decade beforehand in the Crimean War. Being chiefly concerned with chipping away at the clauses of the Paris Peace Treaty of 1856, that excluded it from the Black Sea, it had little interest in joining its former enemies to restore the status quo in Schleswig-Holstein. Besides, being an authoritarian regime, Russia's natural sympathies were with the forces opposed to constitutional principles, hence, with the conservatives in power in Berlin. Matters were equally complicated in relation to Great Britain and France. The two former partners in the Crimean War had fallen out badly, due to Britain suspecting the French leader Emperor Napoleon III (the nephew of Napoleon Bonaparte) of nourishing designs at French world domination. In effect, Britain had resisted all efforts from Napoleon III to redraw the status quo rules to French benefit, which led in turn to France resenting Britain. Knowing this, Bismarck was playing on the hostility of each power toward the other, courting them in turn and promising each support against the other. In relation to Britain, he brandished the stick of a French-sponsored nationalist German Confederation. In relation to France, he dangled the carrot of prospective territorial exchange in the Rhineland.[71]

The dissensions between France and Britain proved impossible to overcome. When Britain resorted to a bluff, by threatening Prussia and Austria with intervention, and actually sending out the order to dispatch the navy to Copenhagen, Bismarck was not impressed. He realized full well that Britain could not have carried out an armed campaign of such a magnitude against two continental powers alone, or at any rate, with just Sweden as an ally. Britain may have been the mistress of the seas, but its army barely exceeded Prussia's. London itself was aware that the only way to carry out the intervention was to bring along France. Napoleon III flatly asserted that he did not trust Britain to honor its commitments which would have left France in a lurch against Prussia and Austria. While it might have been possible for Britain to lure France on its side with concessions, the cost that this would have involved was deemed too large. Accordingly, decision-makers in London started asking themselves if they really wanted a defeat of the German powers, which would have increased French influence and territorial holdings to a scale unseen from the times of the first Napoleon. Better therefore that Denmark be sacrificed than for France to emerge a winner. As a result, Bismarck was allowed to carry the war to completion and coerce Denmark to give up the duchies.[72]

At first, Austria and Prussia each took responsibility for one of the occupied duchies. But while Prussia was bordering Schleswig-Holstein, the possessions of Austria laid far to the south. Thus, Austria would have been content with any deal that would have left Prussia and Austria in a rough balance in Germany, for instance, by exchanging Holstein for a return of a part of Silesia. However, when Bismarck started pressuring Austria to transfer to Prussia absolute control of the duchies, his demand came with the clear implication that Austria was also to accept Prussian supremacy over the states of the German confederation. What Bismarck was promising in return was to support Austria in regaining part of its provinces lost to Italy in 1860, which were not therefore Prussia's to give. It was an unacceptable offer because it was coming with a commensurable Austrian loss of power and status in the German world in favor of Berlin, and Bismarck knew it. His aim was to now provoke Austria into war, but in such a way as to leave it without any great power partner it could rely on for help.

Once again, this was done by exploiting and deepening the rifts between the great powers. There was little hope of support from Russia, which had resented Austria since the Crimean War, when Vienna had sided with Great Britain and France, and which was preoccupied by internal reforms. Bismarck made sure of it by an assiduous courting of St. Petersburg. Bismarck also took steps to gain an alliance with Italy, which sought to gain control of Austrian-held Venice. Meanwhile, after the "meddle and muddle" of the failed Danish intervention, Britain was reluctant to again issue threats it could not carry out. It was also hesitant to take any side between Austria and Prussia, because both states were considered equally guilty of robbing Denmark of Schleswig-Holstein: Britain would not intervene in the division of the spoils. Chief of all, however, Britain remained suspicious of France and of Russia. What Britain sought and what Bismarck promised to deliver was a strong Germany led by Prussia which would act as a counterweight to both powers, and thus represent a de facto dependable partner for Britain on the continent. This left only France. Bismarck bribed Napoleon for his non-involvement in the upcoming conflict with the renewed promise of concessions in Belgium, Luxemburg, or even in the Prussian-owned Rhineland. Bismarck made this clear to the French ambassador, by mentioning that concessions could be offered "anywhere in the world where French was spoken," while also adding that he was "much less German than Prussian." Bismarck stressed to Napoleon that "what Austria can offer him at our expense is more readily obtainable from us." Bismarck was right: the most that Austria could do, was to match Prussia's offer, by also promising France a satellite state in the Rhineland, however on land it did not itself control. The result was that from the point of view of Napoleon, it did not matter who of Prussia and Austria would emerge the winner of a long war that would exhaust both sides. France was going to gain most by staying neutral.[73]

This turned out to be a fatal miscalculation. French neutrality did not benefit Austria, but it allowed Prussia to fight one-on-one against a lone great power, which it won in a matter of only six weeks. The consequences of Austria's overwhelming defeat at Sadowa (Königgrätz) were far-reaching. At Bismarck's insistence, Prussia declined to deprive Austria of its empire—but instead excluded it from Germany. Prussia increased its territory by absorbing not only the duchies, but also several of the Northern German states that had taken Austria's side, while imposing on the rest a Northern German Confederation under its leadership and corralling the southern German states into military alliances. Aside from Austria, the biggest loser from the war was France. In vain did Napoleon III insist on Prussia honoring its pledge to compensate France by supporting its annexation of Luxembourg. Bismarck simply reneged on his former promise, now invoking the excuse that any territory inhabited by Germans was non-negotiable.[74] France's discomfiture was not only restricted to its inability to secure any gains for itself. With two German great powers vying for the control of Germany, France was the strongest power on the continent. But with only one contender and the possibility of German national unification looming, France was confronting the prospect of a more populous, wealthier, stronger rival on its doorstep. Therefore, it was clear that the future, most likely contest was going to take place between France and Prussia for the control of the southern part of Germany, still nominally independent of Prussian control, and implicitly for supremacy on the European continent.

War was also in the interest of Prussia, once it turned out that, far from cowed by Austria's defeat, the southern states rejected Prussian control, voting for anti-Prussian candidates in both state and the Customs Union parliament elections in 1868 and 1869. Hence, Bismarck needed to score a victory over France in order to show the southern states that a future in Germany was their only realistic option left. In the words of his most famous biographer, "Bismarck's goal was ... a crisis with France. He deliberately set sail on a collision course with the intent of provoking either war or a French diplomatic humiliation."[75] It was once again due to Bismarck's skill that in the upcoming confrontation, France was both cast in the wrong and isolated.

The pretext for war occurred in the spring of 1870, when Spain sought to elect a new king. Eventually, the throne was proposed to a Hohenzollern prince. Bismarck strongly campaigned for Prince Leopold to accept the offer, and once he had first declined it, lobbied the Prussian king, William I, to revise his decision, while sending envoys to Madrid to keep the project alive. Bismarck initially sought to achieve a fast victory over France by placing in power a pro-German king in its neighborhood, and then denying having had any knowledge of the event, a "private" decision which involved only the Spanish nation and the prince. As it was, the Hohenzollern candidacy did not stay secret, as Bismarck had hoped, leading to French protests and the Prussian king overriding his chancellor to ask Leopold to withdraw. The French leadership should have been satisfied with that much and let the matter rest, but after the setbacks suffered at Prussian hands since 1866 and under pressure from public opinion, it sought more: to in turn humiliate Prussia. Accordingly, the French ambassador was instructed to again approach the Prussian king and ask him for a further guarantee: that the candidacy would never again be renewed. The king was polite but firm in refusing to offer this assurance. When learning of the meeting, Bismarck edited the dispatch of the interview in such a way that the refusal was made to appear as a brusque dismissal of the French ambassador. The editing was very subtle. Bismarck changed a polite sentence that read the king had "let the ambassador be told through an adjutant that he had now received from the Prince [Leopold] confirmation [about withdrawal of the candidacy] ... and had nothing further to say to the Ambassador" into an inflammatory rebuke: "His Majesty the King had thereupon refused to receive the French Ambassador once more and let him know through an adjutant that His Majesty had nothing further to communicate to the Ambassador." An irate French public then pushed for avenging the offense done to the French representative, which by then had been laid open for the whole world to witness. The French government hurried to declare war, oblivious to the fact that it had been made to look like the villain of the piece. Instead of siding with France, the German states then rallied around the German flag carried by Prussia against what they perceived to be an arrogant French aggressor.[76]

Bismarck was also careful to ensure that France would be deprived of allies when confronting Prussia and its German allies. With French troops still in control of Rome, Italy had limited sympathy for Napoleon III. Bismarck also moved to prevent an alliance between France and Austria. The first measure against this possibility was striking a deal with Russia. In return for Prussian support of rejection of the terms of the Paris Peace Treaty, Russia then mobilized its forces on the borders with Austria. Thus, Russia made sure that Austria stayed quiescent to conflict in the west. But at the same time, Bismarck also moved to patch relations with Austria, now transformed into a dual monarchy by associating Hungary into the ruling of the empire (hence Austria-Hungary) by encouraging it to expand in the east and particularly the Balkans. The consequences of this push were that Austria and Russia's rivalry for the control of the inheritance of provinces controlled by the decaying Ottoman Empire worsened, and both depended on Prussia (and later on Germany) for support against each other. Finally, in regards to Britain, Bismarck sought to embroil London and Paris by playing up the issue of French interest in Belgium, the one portion of Europe that, due to its vicinity to Britain, London could not afford to forsake. Accordingly, from 1866 on, Bismarck encouraged France to take Belgium in compensation for the unification of Germany under Prussian auspices. In fact, a draft treaty to this effect was negotiated in 1866, with Bismarck taking care to have the French ambassador write it down in his own hand. Yet, simultaneously, Bismarck was asking Britain what measures it was prepared to take to defend Belgium from France. When hostilities finally began in 1870, Bismarck promptly leaked the 1866 handwritten draft to the British press to make it seem like a recent French plot to seize Belgium. As a consequence, in the ensuing conflict, Britain favored Prussia over France, not only because it was the latter that started the war, but also because of the belief that a French victory was more to be feared than a Prussian one.[77]

What followed was a repetition of the Austrian scenario, with the bulk of the French forces being decisively defeated six weeks into the war, and with Napoleon III being taken prisoner.[78] While the war dragged on for another nine months, the major remaining obstacle to German unification had been removed. Talk of a united Germany started only days after the crushing victory at Sedan, and the event itself took place in January 1871 in the Hall of Mirrors at Versailles, with the crowning of William I as German Emperor. Bismarck continued in office for another 20 years as chancellor of Germany.

Reassurance and biding your time: the rise of China

In 1978, the year in which Deng Xiaoping unofficially assumed power, China was the least great power in the club. China had been reunified 30 years earlier, and had gone on to fight the US to a standstill in Korea from 1950 to 1953. It even was dubbed in the 1970s a "candidate superpower," too big to be a mere adjunct to either the US or the Soviet Union. But this was flattery. In the wake of decades of political and social upheaval culminating in the Cultural Revolution (1966–1971), in which the country sought to purge lingering capitalist attitudes and thought, China had been left behind. Its GDP ranked tenth in the world, behind not only every other great power, but also second-tier states such as Italy, Canada, Brazil, and Spain. By the onset of the 1980s, China had also been surpassed by India and Mexico, and was outperformed even by its Four Tigers neighbors (Hong Kong, South Korea, Taiwan, and Singapore) in economic growth and use of modern technology. The level of living in China dragged behind not just the capitalist, but also the socialist world. This backwardness also began to show in military equipment and doctrine. In 1969, China was defeated in a border war by the Soviet Union. A decade later it suffered no fewer than 30,000 battle deaths in an effort to discipline Vietnam. As a result, in the 1980s, Japan, not China, was seen by the US as the power to watch in East Asia.[79] Thirty-odd years later, China was barely recognizable. Its share of the world GDP had nearly tripled from 4.9% in 1979 to 13.2%. Basically, the Chinese economy had doubled in size every eight years. As the World Bank put it, China had "experienced the fastest sustained expansion by a major economy in history," averaging 9.6% per year.[80] In 2010, China surpassed Japan in nominal GDP, becoming the second largest economy in the world and prompting widescale recognition that even if it was not yet a superpower in the same league as the US, it was a great power that had broken ahead of the rest of pack.[81] How did China become the current number two great power in less than two generations?

The answer lies in the grand strategy adopted by China. Variously described as "calculative," "camouflaged," and "transitional," this grand strategy was summed up by the pithy 1992 statement by Deng Xiaoping: "*lengjing guancha, chenzho yingfu, wenzhu zehnjiao, taoguang yanghui, yuosuo zuowei,*" or "coolly observe, calmly deal with things, hold one's position, hide our capabilities, bide our time, and accomplish things when possible."[82] Fundamentally, Deng's grand strategy relied on two tactics: consummate pragmatism and masterful reassurance.

Unlike most Chinese leaders of the immediate post-Mao era who either had rarely set foot outside China, let alone in the West, or who viewed the world strictly through the lens of Mao Zedong's thought, Deng had traveled and lived abroad, and realized fully well as a result of his trips to New York and Paris just how far back China had fallen. He also absorbed the lessons of the East Asian economic miracle of the 1960s and 1970s, when Japan and the Four Tigers achieved spectacular levels of development by adopting an export-oriented economic model. Each country had started up by building up labor-intensive industrial sectors, such as textiles, before progressing to capital-intensive industries such as iron and steel, cars, and shipbuilding, and, finally moving to hi-tech industries, such as electronics. Each relied on a disciplined and frugal workforce, on acquiring technical and educational expertise from abroad, as well as on generating the capital needed for development by exporting high-quality goods produced at lower costs than the West. Consequently, Deng embarked on imitating this example.[83]

But what was politically unproblematic in the case of Japan and of the Four Tiger was far harder to do in China because of the constraints of its Communist ideology. Following the East Asian economic model required jettisoning central economic planning, as well as the restrictions and bans on private property, the accumulation of individual wealth, free enterprise, as well as the stigma associated with intellectualism. Essentially, a decision by China to imitate non-communist countries meant acknowledging their superiority from a practical standpoint. What Deng did was to use his authority in the Communist Party to advocate a pragmatic doctrinal re-interpretation, in which scientific, technological, and managerial progress should have been emulated no matter its origin as an asset, not an impediment to building socialism. As Deng put it in 1978:

> for a certain period of time, learning advanced science and technology from the developed countries was criticized as "blindly worshiping foreign things." We have come to understand how stupid this argument is. ... China cannot develop by closing its door, sticking to the beaten track and being self-complacent.

Hence, transforming China into a powerful socialist country by the end of the millennium came before adherence to Marxist-Leninist-Maoist orthodoxy. As Deng was reported as saying: "I do not care about whether the cat is black or white, as long as it catches mice."[84]

As early as 1975, Deng proclaimed China's national interest in pursuing the so-called Four Modernizations in agriculture, industry, military, and science and technology. As soon as he seized power, he dispatched high-level delegations to Japan, as well as Western Europe to investigate ways to improve China's economy and technology, and undertook visits himself to Japan and the US. This was followed by the dispatch of tens of thousands of Chinese students to study abroad on state-sponsored scholarships. But the most notable innovation by Deng was to follow the example of South Korea and Taiwan in setting up four special economic zones (Shenzen, Shantou, Zhuhai, and Xiamen) on the coast opposite Hong Kong, Macao, and Taiwan. The special zones were to attract outside investment through tax incentives and create joint Chinese–foreign ventures for the purpose of manufacturing export products by combining Chinese cheap labor and land and foreign capital, technology, and management techniques. This system was then extended in 1984 to 14 more zones, turning these coastal areas into the engine fueling China's economic growth.[85]

Deng's pragmatism was also in full display in what was to be his "last battle" following the sudden end of the Cold War. From April to June 1989, Beijing's central Tiananmen Square was rocked by demonstrations calling for political reforms, which ended in a violent crackdown ordered by Deng and the other party elders. This was followed by the collapse of Communist regimes in Eastern Europe and the eventual dissolution of the Soviet Union. The lesson that some of the leaders of the Chinese Communist Party extracted from these events was that pro-market reforms had been a mistake that had weakened the Party, and only that a return to stern orthodoxy could save Communism. However, Deng strongly disagreed: that way laid disaster. Watching the summary trial and execution of Romania's Communist dictator Nicolae Ceauşescu, one official remarked that "we will be like this if we don't strengthen proletarian dictatorship." Deng replied: "We will be like this if we don't carry out reforms and don't bring about benefits to the people." Thus, Deng was effectively arguing in favor of a great bargain: as long as the Communist Party ensured a prosperous and strong China, its rule was going to be accepted as legitimate by the Chinese people. But if China closed itself to the world, doom was inevitable. Accordingly, instead of reversing reforms, Deng embarked on a tour of the economic zones in the south, in which he praised reforms, promised catching up to the Four Tigers within 20 years, and lashed out at party conservatives in Beijing. Deng's message was targeted at the Politburo: "whoever is against reform must leave office" and "socialism is not poverty." At age 87, he managed to create a powerful coalition of local officials that eventually prevailed over the top leaders, pushing them to continue the course of reforms under the label "socialism with Chinese characteristics" or "socialist market economy." The result of the southern tour was a second economic opening of China, in which Deng's successors opened the entire country to foreign investment, allowed mass-scale privatization, permitted inefficient state-run enterprises to fail, and endorsed the creation and promotion of domestic brands that could compete internationally. This led to two-digit economic growth from 1993 to the end of the decade.[86]

Pragmatism, however, could not have succeeded without reassurance. This was not a simple matter of, to use Deng's formulas, grasping truth from facts and emancipating one's mind. Since every developed economy was an ally or security partner of Washington, deteriorating relations with the US would have meant not only losing the American market and support, but also jeopardizing ties with Japan, the Four Tigers, the Middle East, not to mention international lending institutions such as the World Bank. In other words, antagonizing the US would have resulted in the interruption of the inflow of foreign capital, technology, education, and expertise without which Deng's modernization agenda could not have been carried out. While chances of conflict were minimal in the late 1970s and early 1980s when the two countries needed each other to counter the Kremlin, the gradual weakening and eventual collapse of the Soviet Union resulted in the gradual worsening of Sino-American relations. By 1991, the US was the lone superpower in the world, unrivalled both militarily and economically, and leading a large section of the international community. Meanwhile, China was outgunned, in the process of developing, and virtually alone. Moreover, the US stood for democracy and human rights, whose spread threatened to overthrow Communist rule, and preserved military and political ties to Taiwan, which China claimed to be part of its territory. Therefore, the risk of Sino-American confrontation was not a matter of speculation. From 1989 to 1999, the US and China clashed on at least four occasions. In 1989, the US and its allies imposed economic sanctions on China that were to last two years for the repression of the Tiananmen demonstrators. In 1993, Washington conditioned the renewal of the most-favored-nation clause to China because of progress in human rights. In 1996, the US sent two air carrier groups in the vicinity of the Taiwan Straits to respond to China's military exercises meant to intimidate the Taiwanese electorate in the island's first ever presidential election. In 1999, in the context of the NATO intervention over Kosovo, the US accidentally bombed the Chinese embassy in Belgrade. Convinced that this was not an accident, China retaliated with massive protests and engaged in a major security debate about whether to revise its relation with the US.[87]

For his part, Deng was not afraid of standing up to Washington. Assuming the worst came to pass and confrontation was forced on it, China's experience throughout the twentieth century had proved that it could withstand foreign invasion or decades of sanctions. This was why Deng argued that China should have kept its calm, observed coolly events as they were occurring, and held firm to its position when vital interests were at stake. China was not to accept being pushed around, whether in relation to human rights, Taiwan, or the retrocession of Hong Kong. But Deng also realized that China provoking or accidentally causing a confrontation with a much more powerful US would have been self-defeating. This was why he recommended that the wisest course of action was reassurance, for China to hide its capabilities and patiently bide its time.[88]

Reassurance was achieved in several ways. First, China downplayed the military factor in its rise, by assigning it a reduced importance and concealing part of its defense spending. Although a nominal part of the Four Modernizations, military modernization took a distinct last place. This marks a significant difference from the self-strengthening of Peter the Great or the Meiji Revolution in which economic and military development occurred at the same time and assumed comparable importance. By contrast, the percentage of total government expenditures China spent on its military after the initiation of reforms was almost halved, going from 15–17% to around 8%. Military personnel, once as high as 4 million, was halved as well. This does not mean that China resigned itself to complete vulnerability. In terms of overall spending, thanks to a booming economy, China ended up spending a good deal more money than it used to on its military. Moreover, the above official Chinese statistics have to be taken in cautiously, since they purposefully leave out a significant part of Chinese military spending on matters such as research and development, paramilitary forces, foreign weapons purchases, and military aid, which taken together would result in a defense budget for the 1990s and 2000s about twice as large as the official version. However, even assuming this is the case, it would not be indicative of a large-scale military build-up, because given its levels of economic growth, China could have, if it had wanted to, spent proportional amounts on its armed forces as had been the case with self-strengthening rising powers that sought to level the playing field with the best militaries in the system. However, as late as 2010, China only accounted for 7.3% of the world's total military spending compared to America's whopping 42.8%.[89]

Second, China supported public relations initiatives as evidence of its peaceful intentions. From the mid-1990s on and especially in the 2000s, China undertook a major diplomatic transformation, securing membership in a host of international political and economic institutions and forums. China became part of the WTO; the Asian Development Bank; the Asia Cooperation Dialogue (comprising 30 members and aiming to become a continental-wide mechanism); the Shanghai Cooperation Organization (China, Russia, Uzbekistan, Kirgizstan, and Kazakhstan); the ASEAN Plus Three (ASEAN and China, South Korea, and Japan); APEC; and the China-ASEAN free trade area, presently the largest free-trade area in the world in terms of population. Beijing also developed economic and diplomatic dialogue ties to African, Arab, and Latino-American states, the EU, and NATO. This was complemented by a soft power initiative in the form of the opening of Confucius institutes, which by 2010 counted no fewer than 322 institutes and 369 classes spread across 96 countries.[90] Furthermore, in 1999, China publicized a new security concept, which stressed the importance of conducting "dialogue, consultation and negotiation on an equal footing" as "the right way to solve disputes and safeguard peace."[91] China then made good on this pledge by reaching peaceful resolutions of territorial disputes with ASEAN members, Russia, Mongolia, Kazakhstan, Kirgizstan, and Tajikistan. In so doing, China frequently gave in to the demands of the other side: for instance, in the case of a dispute with Tajikistan over 28,000 square kilometers, China accepted a settlement that allowed it only 1,000 square kilometers. Similarly, China signed the Declaration on the Conduct of Parties in the South China Sea (2002) proposed by ASEAN, even though the draft, which committed Beijing to resolving disputes in the contested area without resort to the threat of force and through mutual consultations with the other states involved, reflected ASEAN's demands.[92]

Third, China has fought back against suggestions that it harbors any designs on regional or systemic leadership by purposefully maintaining a low international profile. To this end, China has routinely minimized the extent of its economic and, as seen in the above, military capabilities, for instance by reporting levels of comprehensive national power that showed China in the eighth place in the world instead of the third or fourth. China also referred to itself until as late as 2008 as a developing state instead of a rising power or even an important state. This self-censure was evidenced by China's effort in the early 2000s to promote the slogan "peaceful rise," only to replace it with the even blander "peaceful development," so as to calm any misgivings about the security implications of a rising power.[93] Furthermore, China has vehemently denounced the pursuit of hegemony by any power as a former victim of great power expansion—by implication this being something it would refuse to ever contemplate itself. As Prime Minister Wen Jiabao put it in 2004, China's development "will not come at the cost of any other country, will not stand in the way of any other country, nor pose a threat to any other country."[94] Finally, China has sought to discredit the idea of a "China threat," by arguing that that Western analyses that inflate its economic performance and military spending do so on purpose in order to portray Beijing as menacing.[95]

Fourth, China ensured that its rise also benefitted other developed economies so that any actions to curtail it would also hurt its initiators. Thus, Chinese leaders spoke openly of a "win-win diplomacy" and of China's development as an opportunity, not a threat. Trade is considered to account for twice as much for China's economy than for either the US or the EU. This means that China is more integrated than either Washington or Brussels in the global trade system, and while this increases its vulnerability, it also increases its leverage on other economies as a buyer, supplier, or investor. Other developed economies depend on China as an endless reservoir of inexpensive labor and therefore a core product assembly hub; as a growing market of 1.3 billion consumers; and due to its sustained economic growth, a profitable destination for foreign investment. By the 2000s, China accounted for 20–30% of the export growth of Japan, the Four Tigers, and the US, while foreign direct investment represented 35% of the Chinese GDP. The US was China's biggest export destination, ahead of Japan and Hong Kong, and its second largest investor after Japan. Consequently, containing, pressuring, or marginalizing China would have had repercussions for the global economy in terms of loss of profits and investments or losing business opportunities to competitors unwilling to sever their ties to Beijing. Furthermore, due to its favorable trade balance, by the mid-2000s, China had amassed the "largest national hoard in the world" in foreign exchange reserve holdings, a sum hovering around 1.8 trillion dollars, of which an estimated 60–75% has been invested in US treasury bonds, which also gave China a "nuclear option" of sorts of dumping US assets on the market in retaliation for aggressive moves by the US.[96]

Finally, and most importantly, China has stuck to Deng's advice by repeatedly avoiding over-reacting in disputes with the US. To avoid hostility and even violence that would have run havoc for its chances to rise, it was imperative that

China did not respond to what it saw as American provocations, such as interference in the way it ran its domestic affairs; support for Taiwan; endorsement of Japan, Vietnam, and the Philippines in their disputes with China; and interventionism, especially in former Yugoslavia. Even though private criticism of the US interference and "hegemonism" was routinely circulated among Communist elites, it was rarely allowed, with the exception of the bombing of the Kosovo embassy, to be expressed into public anti-American or anti-Western outbursts. The history of Sino-American relations in between 1978 and 2010 has been one of Chinese flexibility, with Beijing offering, albeit discretely, a long series of unmistakable concessions. Under Deng's leadership, China accepted the 1979 Taiwan Relations Act, which allowed the US to continue low-level diplomatic ties to the island, as well as to sell it weapons; it put up with major US arms sales to Taipei in 1980 and 1990; and refrained from lashing at the West for the sanctions imposed in the aftermath of the Tiananmen crackdown. Deng's successors saw as well the wisdom of reassurance after blundering in the Taiwan Straits crisis of 1995–1996, in which their attempts at coercive diplomacy against Taiwan ended up by provoking American intervention and a strengthening of the US–Japanese alliance. The lesson extracted from the crisis and reaffirmed in the great security debate of 1999 was that the US's dominant influence in world politics was likely to persist for decades. Consequently, as long as an unbridgeable gap of power persisted between them, China should have refrained from recklessly challenging the US. As a result, China did not attempt to form a great power coalition to reign in a dangerously powerful, aggressive, and unilateralist US. Instead, China sought a "strategic partnership" with the US in the late 1990s; rallied behind Washington in the war on terror in the fall of 2001 by helping convince Pakistan to support military action against the Taliban; endorsed sanctions against North Korea in 2006, 2009, and 2010; and maintained a low profile in relation to the US invasion and later occupation of Iraq.[97]

This complex, multilayered, reassurance grand strategy has proved a considerable success, convincing the world that China has become, in the words of US Secretary Robert Zoellick a "responsible stakeholder in the international system."[98] Whether this transformation is genuine or cosmetic is beside the point. As Deng would have pointed out, results are what matter. In this respect, China has won crucial time for its economy not only to catch up, but also to surpass most of its competitors and be able to face the international environment with increased confidence.

Notes

1 Parker, "Western Way of Warfare," 2–12, 6–7. For the argument that the conduct of warfare transforms the state domestically see Hintze, "Military Organization," 178–215; Tilly, "Reflections on the History of European State-Making," 42.
2 Resende-Santos, "Anarchy and Emulation of Military Systems," 193–260, 207–8; Horowitz, *Diffusion of Military Power*. For internal balancing and emulation, see Kenneth Waltz, *Theory of International Politics* (Reading: Addison Wesley, 1979), 124–7, 168; Taliaferro, "Neoclassical Realism and Resource Extraction," 194–226.

3 Jansen, *Making of Modern Japan*; Duus, *Abacus and Sword*; Paine, *Japanese Empire*.
4 Japan on the eve of the Meiji revolution had a population of 34 million, larger than Great Britain's, on par with Austria-Hungary's, and just 4 million shy of those of the US and France. *National Material Capabilities v5.0* at http://www.correlatesofwar.org /data-sets/national-material-capabilities.
5 For the idea of the rising power as an opportunistic aggressor, not a mindless one, see Mearsheimer, *Tragedy*, 37.
6 For a study that equivalate windows of opportunity with power shifts see Van Evera, *Causes of War*, 73–5. For a critique that warns that decision-makers do not react automatically or unproblematically to shifts in relative power see Lebow, "Windows of Opportunity."
7 George, "Strategies for Crisis Management," 382–3.
8 For a recent similar argument, see Shifrinson, *Rising Titans, Falling Giants*.
9 As Van Evera writes, "a resource is cumulative if its possession helps its possessor to protect or acquire other resources." Van Evera, *Causes of War*, 105; Kamen, *Spain's Road to Empire*.
10 Van Evera, *Causes of War*, 107–12; Kaysen, "Is War Obsolete?"
11 Brooks, "Globalization of Production." For the opposite view that conquest may still pay as long as the occupying power is willing to act as a ruthless invader by using coercion to extract benefits from the vanquished in the form of tribute or plundering resources, see Liberman, *Does Conquest Pay?*
12 Blainey, *Causes of War*, 57–67.
13 Jervis, *System Effects*, chap. 5; Schweller, *Deadly Imbalances*, 48; Hui, *War and State*, 27–8, chap. 2; Simmel, *Sociology*, 154–69.
14 Snyder, *Alliance Politics*, 318–20. Another form of divide and conquer is to sow division not just between great powers, but also between various domestic factions inside individual great powers, complicating their ability to mobilize and extract resources from the society and to reach effective decisions. Nexon, "Balance of Power in the Balance," 345.
15 Jervis, *System Effects*, 178–83; Crawford, "Preventing Enemy Coalitions," 160–4, 180–5.
16 The position of the rising power appears very much like that of the balancer or the state holding the balance of power, not strong enough to be itself a main contender, but strong enough to determine the outcome of the contest. However, unlike the balancer, the rising power does not need to exert decisive influence on who wins or loses. It is enough for it to make a difference in the contest conferring the overall advantage to one side. Great powers agonized over the allegiance of Italy prior to World War I, and the superpowers engaged in frequent bidding contests for the support of states such as Egypt, Cuba, Ethiopia, and China, which could not have decided the result of the Cold War. Jervis, *System Effects*, 181–2.
17 Ibid., 177–8, 183, 185, 188 fn. 42.
18 Bertie, "'Private and Secret' Memorandum, 27 October 1901," 464–9.
19 Glosny, *Grand Strategies*, 46.
20 Kydd, "Sheep in Sheep's Clothing," 140–5; Glaser, *Rational Theory*, 64–8; Jervis, "Cooperation Under the Security Dilemma." The rising power does not need to disarm fully or to renounce any offensive capabilities to convey its benign intentions, but rather to forgo taking advantage of opportunities to increase its aggressive potential.
21 Johnston, *Social States*; Ikenberry, *After Victory*.
22 Nye, *Soft Power*; Kurlantzick, *Charm Offensive*.
23 Copeland, *Economic Interdependence and War*; Edelstein, *Over the Horizon*, 15–17, 23–4.
24 Levy, "Declining Power," 96; Powell, "War as a Commitment Problem," 188.
25 MacMillan, *War That Ended Peace*, 94–5.

26 Tompson, *Khrushchev*, 171.
27 Levy, "Declining Power," 96.
28 Lebow, *Why Nations Fight*, 93–4.
29 Anderson, *Peter the Great*, 1–13; Stevens, *Russia's Wars of Emergence.*
30 LeDonne argues in favor of the existence of a Russian long-term grand strategy of seeking hegemony in the Heartland, the area between the Baltic and the Caspian seas. This interpretation also coincides with the alleged political testament of Peter the Great, which councils Russia to divide Poland, intervene in Germany, become the strongest power in the Baltics and the Black Sea, and eventually aim at European and global domination. However, there is no proof that Peter was aiming for supremacy— rather, given the weakness of Russia, he merely sought recognition as one of the great powers. Furthermore, he prioritized reforms and position in Europe over expansion at the expense of the Ottoman Empire. The testament itself was proven to have been a forgery, most likely orchestrated by French propaganda in the context of Napoleon's 1812 campaign against Russia. LeDonne, *Grand Strategy of the Russian Empire*, 6–7, 85–92; Lehovich, "The Testament of Peter the Great."
31 Hughes, *Peter the Great*, 9–26.
32 Ibid., 40–51.
33 Hughes, *Russia in the Age of Peter the Great*, 385–9.
34 Ibid., 65–9, 89–91; Anderson, *Peter the Great*, 99–101.
35 Ibid., 104, 113–19, 132–6, 145–56, Hughes, *Russia in the Age of Peter the Great*, 145–54.
36 Ibid., 135–42; Fuller, *Strategy and Power*, 57–9.
37 Anderson, *Peter the Great*, 102–3.
38 Hughes, *Russia in the Age of Peter the Great*, 384; also see Bushkovitch, *Peter the Great.*
39 Peter never completely gave up enlargement in the south. In 1711, he launched an offensive at the head of a host of 40,000 men, hoping for a general uprising of the Orthodox Christian population in the Ottoman Empire; and the closing years of his reign witnessed a renewed attempt at conquest against Persia. But expansion in the south was clearly subordinated to interests in the west. The best evidence was Peter's readiness to make concessions in the south in order to conserve gains secured from Sweden. When the campaign of 1711 ended in disaster, with the tsar's forces surrounded by an Ottoman army three times its size at the Pruth River in Moldavia (present day Romania), he agreed to return to the Ottomans the fortress of Azov, which he had seized in 1796, and, an even greater personal sacrifice, to dismantle Russia's fleet he had built there.
40 Anderson, *Peter the Great*, 50.
41 Hughes, *Peter the Great*, 195.
42 Fuller, *Strategy and Power*, 38–41.
43 Ibid., 83, 71–8.
44 A further mistake by Charles XII was to count on the hypothetical intervention on his side by the Ottoman Empire and the Crimea Khanate, which never materialized, as well as on a large contribution of 20,000 riders from the Zaporizhian Cossacks, which turned in practice to be ten times smaller.
45 Hughes, *Peter the Great*, 78–84; Fuller, *Strategy and Power*, 42–3, 79–82.
46 Hughes, *Russia in the Age of Peter the Great*, 50–7.
47 Anderson, *Peter the Great*, 83–6. See for a dissenting argument, which makes Russia's acceptance as a great power conditional not only on capabilities and success, but also on cultural affinities, Neumann, *Uses of the Other.*
48 Scott, "Prussia's Emergence," 152–76, 152–4; Scott, "1763–1786," 177–200, 186.
49 Frederick William I was one of the innovators of the reserve system—with all able-bodied men of serving age being trained for two years and then returning for exercises

two months every year. The rest of the time, the soldier, technically on leave, was free to return home and pursue his peacetime profession, costing the state no expense and adding to its productivity.

50 Schulze, "Prussian Military State," 201–19, 201, 205–6; Blanning, *Frederick the Great*, 14–15, 288–91; Showalter, *Wars of Frederick the Great*, 18–24.

51 Scott, "Prussia's Emergence," 156; Blanning, *Frederick the Great*, 127–8; Ritter, *Frederick the Great*, 21.

52 As early as 1731, Frederick was busy listing territories that would have made attractive acquisitions for Prussia. Blanning, *Frederick the Great*, 85. This does not mean that he was a warmonger at heart. The same year he invaded Silesia he published a treatise refuting Machiavelli, which vehemently condemned selfish wars of conquest. But Frederick was ultimately a realist dealing with conditions as they were, not as he might have wished them to be. As he noted, "at all times, it was the principle of great states to subjugate all whom they could and to extend their power continuously." Ritter, *Frederick the Great*, 66, 64–8, 32–3.

53 Blanning, *Frederick the Great*, 82–3, 87–97.

54 Ibid., 98; Friederich des Grossen, "Die Politischen Testamente," at https://archive.org/stream/diepolitischente00freduoft/diepolitischente00freduoft_djvu.txt, 59.

55 Scott, "Prussia's Emergence," 159–60; Blanning, *Frederick the Great*, 98–100, 199–200; Clark, *Iron Kingdom*, 190–2; Ritter, *Frederick the Great*, 76–82.

56 Friederich des Grossen, "Die Politischen Testamente," 49, 50, 62, 64. As it happened, Frederick lived long enough to orchestrate the First Partition of Poland in 1772, by which Prussia gained most of Polish West Prussia, finally joining East Prussia to Brandenburg.

57 Blanning, *Frederick the Great*, 102–26; Showalter, *Wars of Frederick the Great*, chap. 2; Clark, *Iron Kingdom*, 195–6.

58 In Frederick's self-interested reasoning, France would have had no reason to complain since in the eventuality of a war, it was free to attack with Prussian help the Austrian Netherlands (Belgium), instead of Hanover, and would thus gain a far richer prize. Of course, doing so would have turned France against Prussia's chief opponent Austria, while being of no use for the purpose of pressuring France's own chief enemy, Britain.

59 Blanning, *Frederick the Great*, 198–218; Clark, *Iron Kingdom*, 196–200.

60 Scott, "Prussia's Emergence," 168–9, 173–4; Blanning, *Frederick the Great*, 222, 268, also see chap. 10; Clark, *Iron Kingdom*, 200–1.

61 The most detailed account is Showalter, *Wars of Frederick the Great*, chaps. 4–6.

62 Blanning, *Frederick the Great*, 272–9.

63 Clark, *Iron Kingdom*, 510. For Prussia's record of setbacks and humiliations in the late 1840s and 1850s see ibid. 486–97.

64 Kennedy, *Rise and Fall of Great Powers*, 154, 171; Hinsley, *Power and Pursuit of Peace*, 301.

65 Kissinger, *Diplomacy*, 137–8.

66 Steinberg, *Bismarck*, 174, 180–1; Gall, *Bismarck*, 99, 204.

67 Bismarck, "Memorandum Dictated at Kissingen," 54–5.

68 Bismarck's strategy was more than pactomania, a continual expansion of a network of alliances and alignments. The network was efficient primarily because the differences between Berlin's partners were so considerable and so magnified due to Berlin's divisive diplomacy that they could not contemplate alternative alignments—as such they remained dependent on Bismarck's support. Joffe, "Bismarck or Britain?"

69 As Lord Salisbury put it of Bismarck: "I believe he is still true to the main principle of his policy—employing his neighbors to pull each other's teeth." Kennedy, *Rise of the Anglo-German Antagonism*, 194–5.

70 Langer, *European Alliances and Alignments*, 197; Pflanze, *Bismarck* vol. 2, 253, 501; Bismarck, "Memorandum Dictated at Kissingen," 54–5.

71 Mosse, *European Powers*, 110, 176–7, 211–2, 221–2, 298–9; Pflanze, *Bismarck* vol. 1 (Princeton: Princeton University Press, 1990), 243–50; Steefel, *Schleswig-Holstein*, chaps. 3–4.

72 Stacy Goddard has argued that Bismarck was relying on a "legitimation strategy" by making rhetorical appeals to the principles professed by each side, respectively respect for international treaties for Britain, nationalism for France, and conservativism for Russia, thus placing each power in the awkward position of having to trample on its cherished beliefs in order to check Prussia. Goddard, "When Right Makes Might." However, Goddard's interpretation suffers from two flaws. First, it ignores that Prussia set aside any effort at legitimation once the war was pursued into Denmark proper. Steefel comments that, by the time the hostilities were initiated, Austria's and Prussia's assurances that their actions were not meant to violate the treaty and hence the integrity of Denmark were worth little more than the paper on which they were written. Yet, despite the discrepancy between words and deeds, the other powers did not act to check the two aggressors. Steefel, *Schleswig-Holstein*, 168. Second, the same rhetorical trick cannot account for the continued great power tolerance toward Prussia in its subsequent wars with Austria and then France, in which they should have been aware of their attempted manipulation by Bismarck. Ibid., 180–4, 190–8.

73 Pflanze, *Bismarck* vol. 1, 264–5, 298–303; Mosse, *European Powers*, 221–4, 231–2, 236–7; Pottinger, *Napoleon III*, 10, 21, 27–32, 70–2, 82–4, 136–43, 146–7; Clark, *Iron Kingdom*, 533–4; Millman, *British Foreign Policy*, 35–7.

74 Pflanze, *Bismarck*, vol. 1, 371–81.

75 Pottinger, *Napoleon III*, 165–6; Carr, *Origins of Wars of German Unification*, 171–3; Böhme, *Foundation of German Empire*), 196–9; Pflanze, *Bismarck* vol. 1, 405–8, 451, 462.

76 Steinberg, *Bismarck*, 283–9; Pflanze, *Bismarck* vol. 1, 462–9; Steefel, *Bismarck*, 56–88; Wetzel, *Duel of Giants*. It is true that Prussia was resented for its 1866 victory; but France was seen as a hereditary enemy of Germany, due to Napoleon I's aggressive policies and to the French support for Italy against Austria in the 1859 war. The tiebreaker in choosing between these two otherwise disliked powers was that it was France that appeared in the posture of aggressor. By its pretensions, even after the Spanish candidacy had been withdrawn, for a further guarantee from King Wilhelm, France was seen as having behaved in an insulting manner not only toward Prussia and its king, but also toward the German nation as a whole.

77 Pflanze, *Bismarck* vol. 1, 437–45, 472–3; Millman, *British Foreign Policy*, 65–6, 119–21, 200–1; Mosse, *European Powers*, 306–13.

78 Howard, *Franco-Prussian War*.

79 Maddison, *Chinese Economic Performance*; Garver, *China's Quest*, 678–9.

80 Maddison, *Chinese Economic Performance*, 103; International Monetary Fund, *World Economic Outlook Database*, April 2010 at https://www.imf.org/external/pubs/ft/weo/2010/01/weodata; Wayne Morrison, "China's Economic Rise: History, Trends, Challenges, and Implications for the United States," available at https://fas.org/sgp/crs/row/RL33534.pdf.; Holz, "China's Economic Growth."

81 "China: Second in Line," *The Economist*, August 16, 2010; Brooks and Wohlforth, "Rise and Fall of Great Powers," 15, 32.

82 Pillsbury, *China*, xxxix; Feng, "China's Rise Will Be Peaceful," 34–54, 46–7; also see Swaine and Tellis, *Interpreting China's Grand Strategy*, 112–4; Goldstein, *Rising to the Challenge*, 38–48; Goldstein, "Emerging China's Emerging Grand Strategy," 57–106; Lampton, *Three Faces of Chinese Power*, 25–32.

83 Garver, *China's Quest*, 358, 676–8.

84 Ibid., 349–51; Vogel, *Deng Xiaoping*, chaps. 7, 10–12.

85 Garver, *China's Quest*, 355–77; Vogel, *Deng Xiaoping*, 217–27, 394–411, 418–21, 455–64.

86 Garver, *China's Quest*, 483–4, 512; Baum, *Burying Mao*, 313–40; Vogel, *Deng Xiaoping*, 664–5, 669–84.
87 Goldstein, *Rising to the Challenge*, 12; Mann, *About Face*, chaps. 12 and 15; Garver, *Face Off*; Finkelstein, *China Reconsiders Its National Security*.
88 Garver, *China's Quest*, 421–2, 484. In fact, Deng went so far as to threaten military action against Great Britain in case Hong Kong was not restored to Chinese sovereignty on the lease's date of expiration in 1997.
89 Shambaugh, *Modernizing China's Military*, 189–91 193, 215–22; Goldstein, *Rising to the Challenge*, 54–6; Stockholm Peace Research Institute at http://www.sipri.org/research/armaments/milex/factsheet2010.
90 Goldstein, *Rising to the Challenge*, 119–27, 130–5; Medeiros and Fravel, "China's New Diplomacy"; Christensen, "Fostering Stability or Creating a Monster"; Kurlantzick, *Charm Offensive*, chap. 5; Lampton, *Three Faces of Chinese Power*, 155–6; "Confucius Institutes in the World," at http://college.chinese.cn/en/node_1979.htm.
91 Finkelstein, "China's New Security Concept."
92 Fravel, "Regime Insecurity and International Cooperation"; Fravel and Medeiros, "China's New Diplomacy." It should be noted, however, that some of these disputes, such as the one concerning the Spratly Islands with Vietnam and the Philippines resurfaced in the 2010s. Swaine, *America's Challenge*.
93 Pilsbury, *China*, 218–42; Lampton, *Three Faces of Chinese Power*, 25, 33.
94 Goldstein, *Rising to the Challenge*, 114–15; Kurlantzick, *Charm Offensive*, 38.
95 Deng, "Reputation and the Security Dilemma," 186–214, 199–201; Callahan, "How to Understand China."
96 Lampton, *Three Faces of Chinese Power*, 88–116, esp. 114–16; Bergsten, Gill, Lardy, and Mitchell, *China*.
97 Garver, *China's Quest*, 404–8, 419–26, 483–4, 498–502, 634–6 637–44, 651–3; Finkelstein, *China Reconsiders Its National Security*.
98 Robert Zoellick, "Whither China: From Membership to Responsibility," September 21, 2005 at https://2001–2009.state.gov/s/d/former/zoellick/rem/53682.htm.

Bibliography

Anderson, MS. *Peter the Great*. New York: Longman, 1995.
Baum, Richard. *Burying Mao: Chinese Politics in the Era of Deng Xiaoping*. Princeton: Princeton University Press, 1994.
Bergsten, Fred, Bates Gill, Nicholas Lardy, and Derek Mitchell. *China: The Balance Sheet: What the World Needs to Know Now About the Emerging Superpower*. New York: Public Affairs, 2006.
Bertie, Francis. "'Private and Secret' Memorandum, 27 October 1901." In *The Foreign Policy of Victorian Britain, 1830–1902*, edited by Kenneth Bourne. Oxford: Clarendon, 1970.
Bismarck, Otto von. "Memorandum Dictated at Kissingen, June 15th, 1877." In *German Diplomatic Documents, 1870–1914*, edited by ETS Dugdale, vol. I. London: Methuen, 1928.
Blainey, Geoffrey. *The Causes of War*. 3rd edn. New York: Free Press, 1988.
Blanning, Tim. *Frederick the Great: King of Prussia*. New York: Random House, 2016.
Böhme, Helmut. *The Foundation of the German Empire: Select Documents*. Oxford: Oxford University Press, 1971.
Brooks, Stephen. "The Globalization of Production and the Changing Benefits of Conquest." *Journal of Conflict Resolution* 43 (October 1999): 646–70.
Brooks, Stephen and William Wohlforth. "The Rise and Fall of the Great Powers in the Twentieth First Century: China's Rise and the Fate of America's Global Position." *International Security* 40 (Winter 2015): 7–53.

Bushkovitch, Paul. *Peter the Great: The Struggle for Power.* Cambridge: Cambridge University Press, 2001.

Callahan, William. "How to Understand China: The Dangers and Opportunities of Being a Rising Power." *Review of International Studies* 31 (October 2005): 701–14.

Carr, William. *The Origins of the Wars of German Unification.* London: Longman, 1991.

Christensen, Thomas. "Fostering Stability or Creating a Monster." *International Security* 31 (Summer 2006): 81–126.

Clark, Christopher. *Iron Kingdom: The Rise and Downfall of Prussia, 1600–1947.* Cambridge: Harvard University Press, 2006.

Copeland, Dale. *Economic Interdependence and War.* Princeton: Princeton University Press, 2015.

Crawford, Timothy. "Preventing Enemy Coalitions: How Wedge Strategies Shape Power Politics." *International Security* 35 (Spring 2011): 155–89.

Deng, Yong. "Reputation and the Security Dilemma: China Reacts to the China Threat Theory." In *New Directions in the Study of China's Foreign Policy*, edited by Alastair Iain Johnston and Robert Ross. Stanford: Stanford University Press, 2006.

Duus, Peter. *The Abacus and the Sword: The Japanese Penetration of Korea.* Berkeley: University of California Press, 1995.

Edelstein, David. *Over the Horizon: Time, Uncertainty, and the Rise of Great Powers.* Ithaca: Cornell University Press, 2017.

Feng, Zhu. "China's Rise Will Be Peaceful." In *China's Ascent: Power Security and the Future of International Politics*, edited by Robert Ross and Zhu Feng. Ithaca: Cornell University Press, 2008.

Finkelstein, David. *China Reconsiders Its National Security: "The Great Peace and Development Debate of 1999."* Alexandria: CAN Corporation, 2000.

Finkelstein, David. "China's New Security Concept: Reading Between the Lines." *Washington Journal of Modern China* 5 (Spring 1999): 37–49.

Fravel, Taylor. "Regime Insecurity and International Cooperation: Explaining China's Compromises in Territorial Disputes." *International Security* 30 (Fall 2005): 47–83.

Fuller, William. *Strategy and Power in Russia, 1600–1914.* New York: Free Press, 1992.

Gall, Lothar. *Bismarck: The White Revolutionary*, vol. 1. London: Unwyn Hyman, 1986.

Garver, John. *China's Quest: The History of the Foreign Relations of the People's Republic of China.* Oxford: Oxford University Press, 2016.

Garver, John. *Face Off: China, the United States, and Taiwan's Democratization.* Seattle: University of Washington Press, 1997.

George, Alexander. "Strategies for Crisis Management." In *Avoiding War: Problems of Crisis Management*, edited by Alexander George. Boulder: Westview Press, 1991.

Glaser, Charles. *Rational Theory of International Politics: The Logic of Competition and Cooperation.* Princeton: Princeton University Press, 2010.

Glosny, Michael. *Grand Strategies of Rising Powers: Reassurance, Coercion, and Balancing Responses.* Unpublished Ph.D. dissertation, MIT, 2012.

Goddard, Stacy. "When Right Makes Might: How Prussia Overturned the European Balance of Power." *International Security* 33 (Winter 2008): 110–42.

Goldstein, Avery. "An Emerging China's Emerging Grand Strategy: A Neo-Bismarckian Turn?" In *International Relations Theory and the Asia Pacific*, edited by G John Ikenberry and Michael Mastanduno. New York: Columbia University Press, 2003.

Goldstein, Avery. *Rising to the Challenge: China's Grand Strategy and International Security.* Stanford: Stanford University Press, 2005.

Hinsley, FH. *Power and the Pursuit of Peace: Theory and Practice in the History of Relations Between States*. Cambridge: Cambridge University Press, 1963.

Hintze, Otto. "Military Organization and the Organization of the State." In *The Historical Essays of Otto Hintze*, edited by Felix Gilbert. New York: Oxford, 1975.

Holz, Carsten. "China's Economic Growth 1978–2025: What We Know Today About China's Economic Growth Tomorrow." *World Development* 36, no. 10 (2008): 1665–91.

Horowitz, Matthew. *The Diffusion of Military Power: Causes and Consequences for International Politics*. Princeton: Princeton University Press, 2010.

Howard, Michael. *The Franco-Prussian War: The German Invasion of France, 1870–1871*. London: Rupert Hart-Davis, 1962.

Hughes, Lindsey. *Peter the Great: A Biography*. New Haven: Yale University Press, 2004.

Hughes, Lindsey. *Russia in the Age of Peter the Great*. New Haven: Yale University Press, 2000.

Hui, Victoria Tin-bor. *War and State Formation in Ancient China and Early Modern Europe*. New York: Cambridge University Press, 2005.

Ikenberry, John G. *After Victory: Institutions, Strategic Restraint, and the Rebuilding of Order After Major Wars*. Princeton: Princeton University Press, 2001.

Jansen, Marius. *The Making of Modern Japan*. Cambridge: Harvard University Press, 2002.

Jervis, Robert. "Cooperation Under the Security Dilemma." *World Politics* 30 (January 1978): 167–214.

Jervis, Robert. *System Effects: Complexity in Politics and Social Life*. Princeton: Princeton University Press, 1997.

Joffe, Josef. "Bismarck or Britain? Toward an American Grand Strategy After Bipolarity." *International Security* 19 (Spring 1995): 94–117.

Johnston, Alastair Iain. *Social States: China in International Institutions, 1980–2000*. Princeton: Princeton University Press, 2008.

Kamen, Henry. *Spain's Road to Empire: The Making of a World's Power 1492–1763*. London: Penguin Books, 2002.

Kaysen, Carl. "Is War Obsolete? A Review Essay." *International Security* 14 (Spring 1990): 42–64.

Kennedy, Paul. *The Rise and Fall of Great Powers: Economic Change and Military Conflict from 1500 to 2000*. New York: Random House, 1987.

Kennedy, Paul. *The Rise of the Anglo-German Antagonism, 1860–1914*. Boston: Allen & Unwyn, 1980.

Kissinger, Henry. *Diplomacy*. New York: Simon & Schuster, 1994.

Kurlantzick, Joshua. *Charm Offensive: How China's Soft Power Is Transforming the World*. New Haven: Yale University Press, 2007.

Kydd, Andrew. "Sheep in Sheep's Clothing: Why Security Seekers Do Not Fight Each Other." *Security Studies* 7, no. 1 (1997): 114–55.

Lampton, David. *The Three Faces of Chinese Power: Might, Money, and Minds*. Berkeley: University of California Press, 2008.

Langer, William. *European Alliances and Alignments, 1871–1880*. New York: Knopf, 1950.

Lebow, Richard Ned. *Why Nations Fight: Past and Future Motives for War*. Cambridge: Cambridge University Press, 2010.

Lebow, Richard Ned. "Windows of Opportunity: Do States Jump Through Them?" *International Security* 9 (Summer 1984): 147–86.

LeDonne, John. *The Grand Strategy of the Russian Empire, 1650–1831*. New York: Oxford University Press, 2004.

Lehovich, Dimitry. "The Testament of Peter the Great." *The American Slavic and East European Review* 7 (April 1948): 111–24.

Levy, Jack. "Declining Power and the Preventive Motivation for War." *World Politics* 40 (October 1987): 82–107.

Liberman, Peter. *Does Conquest Pay? The Exploitation of Occupied Industrial Societies*. Princeton: Princeton University Press, 1996.

MacMillan, Margaret. *The War that Ended Peace: The Road to 1914*. New York: Random House, 2013.

Maddison, Angus. *Chinese Economic Performance in the Long Run*. 2nd edn. Paris: OECD Publications, 2007.

Mann, James. *About Face: A History of America's Curious Relation with China from Nixon to Clinton*. New York: Vintage Books, 2000.

Mearsheimer, John. *Tragedy of Great Power Politics*. New York: W. W. Norton, 2001.

Medeiros, Evan and Taylor Fravel, "China's New Diplomacy." *Foreign Affairs* 82 (November 2003): 22–35.

Millman, Richard. *British Foreign Policy and the Coming of the Franco-Prussian War*. Oxford: Clarendon, 1965.

Mosse, WE. *The European Powers and the German Question, with Special Reference to England and Russia*. Cambridge: Cambridge University Press, 1958.

Neumann, Iver. *Uses of the Other: The 'East' in European Identity Formation*. Minneapolis: University of Minnesota Press, 1999.

Nexon, Daniel. "The Balance of Power in the Balance." *World Politics* 61 (April 2009): 330–59.

Nye, Joseph. *Soft Power: The Means to Success in World Politics*. New York: Public Affairs, 2004.

Paine, SCM. *The Japanese Empire: Grand Strategy from the Meiji Restoration to the Pacific War*. Cambridge: Cambridge University Press, 2017.

Parker, Geoffrey. "The Western Way of Warfare." In *The Cambridge Illustrated History of Warfare: The Triumph of the West*, edited by Geoffrey Parker. Cambridge: Cambridge University Press, 2002.

Pflanze, Otto. *Bismarck and the Development of Germany: The Period of Consolidation, 1871–1880*, vol. 2. Princeton: Princeton University Press, 1990.

Pflanze, Otto. *Bismarck and the Development of Germany: The Period of Unification, 1815–1871*, vol. 1. Princeton: Princeton University Press, 1990.

Pillsbury, Michael. *China Debates the Security Environment*. Washington: National Defense University Press, 2000.

Pottinger, Ann. *Napoleon III and the German Crisis*. Cambridge: Harvard University Press, 1966.

Powell, Robert. "War as a Commitment Problem." *International Organization* 60 (Winter 2006), 169–203.

Resende-Santos, Joao. "Anarchy and the Emulation of Military Systems: Military Organization and Technology in South America, 1870–1930." In *Realism: Restatements and Renewal*, edited by Benjamin Frankel. London: Frank Cass, 1996.

Ritter, Gerhard. *Frederick the Great: A Historical Profile*. Berkeley: University of California Press, 1968.

Schulze, Hagen. "The Prussian Military State, 1763–1806." In *The Rise of Prussia, 1700–1830*, edited by Philip Dwyer. New York: Pearson, 2000.

Schweller, Randall. *Deadly Imbalances: Tripolarity and Hitler's Strategy of World Conquest*. New York: Columbia University Press, 1998.

Scott, HM. "1763–1786: The Second Reign of Frederick the Great." In *The Rise of Prussia, 1700–1830*, edited by Philip Dwyer. New York: Pearson, 2000.

Scott, HM. "Prussia's Emergence as a European Great Power, 1740–1763." In *The Rise of Prussia, 1700–1830*, edited by Philip Dwyer. New York: Pearson, 2000.

Shambaugh, David. *Modernizing China's Military: Progress, Problems, and Prospects*. Berkeley: University of California Press, 2002.

Shifrinson, Itzkowitz Joshua. *Rising Titans, Falling Giants: How Great Powers Exploit Power Shifts*. Ithaca: Cornell University Press, 2018.

Showalter, Dennis. *The Wars of Frederick the Great*. New York: Longman, 1996.

Simmel, Georg. *The Sociology of Georg Simmel*, translated by Kurt Wolff. New York: Free Press, 1964.

Snyder, Glenn. *Alliance Politics*. Ithaca: Cornell University Press, 1997.

Steefel, Lawrence. *Bismarck, the Hohenzollern Candidacy, and the Origins of the Franco-German War of 1870*. Cambridge: Harvard University Press, 1962.

Steefel, Lawrence. *The Schleswig-Holstein Question*. Cambridge: Harvard University Press, 1932.

Steinberg, Jonathan. *Bismarck: A Life*. New York: Oxford University Press, 2011.

Stevens, Carol. *Russia's Wars of Emergence, 1460–1730*. London: Routledge, 2007.

Swaine, Michael. *America's Challenge: Engaging a Rising China in the Twenty-First Century*. Washington: Carnegie Endowment for International Peace, 2011.

Swaine, Michael and Ashley Tellis. *Interpreting China's Grand Strategy: Past, Present, and Future*. Santa Monica: Rand, 2000.

Taliaferro, Jeffrey. "Neoclassical Realism and Resource Extraction: State Building for Future War." In *Neoclassical Realism, the State, and Foreign Policy*, edited by Steven Lobell, Norrin Ripsman, and Jeffrey Taliaferro. New York: Cambridge University Press, 2009.

Tilly, Charles. "Reflections on the History of European State-making." In *The Formation of National States in Western Europe*, edited by Charles Tilly. Princeton: Princeton University Press, 1975.

Tompson, William J. *Khrushchev: A Political Life*. New York: St. Martin's, 1995.

Van Evera, Stephen. *Causes of War*. Ithaca: Cornell University Press, 1999.

Vogel, Ezra. *Deng Xiaoping and the Transformation of China*. Cambridge: Harvard University Press, 2011.

Waltz, Kenneth. *Theory of International Politics*. Reading: Addison Wesley, 1979.

Wetzel, David. *A Duel of Giants: Bismarck, Napoleon III, and the Origins of the Franco-Prussian War*. Madison: University of Wisconsin Press, 2001.

4   The grand strategies of status quo powers I

From 1648 to the present, great powers have relied on several possible grand strategies of status quo: 1) primacy, 2) concert, 3) semi-detachment, and 4) containment. This chapter lays down the tenets of the first two grand strategies before providing appropriate historical illustrations.

Primacy

Primacy is the grand strategy of cementing in place the great power's position as the international number one by taking steps to enhance its lead on the other great powers. While a dominant state may not face any competitors in the present, it has no guarantee that this situation will not change to its detriment in the future. The more its existing circumstances are gratifying, the more likely that what is yet to come can only be worse by comparison and, therefore, must be guarded against.[1] Thus, the dominant state takes advantage of the opportunity created by its moment of supremacy to achieve further moderate gains that help it prevent or manage an upcoming situation where the tables turn against it. The ideal goal of primacy is to make the dominant state's position unassailable, but not all-powerful. In this sense, primacy is a grand strategy of improving upon the status quo, but without upsetting it, which is why it might be understood as status quo plus.[2] This grand strategy can be seen in the cases of Louis XIV's France and, arguably, of the post-Cold War US, as shown in Chapter Seven.

There are several ways in which a dominant state can reinforce its position. The simplest is to field military forces whose total number equals or slightly exceeds the joint forces of the next-in-line powers, so that even a combination of them would not be able to prevail against it. A variation in this policy, especially in present settings where advanced technology determines the strength of a military, is to spend more on the military than the total sum spent by the next-in-line powers. Effectively, the US accounted in 1999 for a military budget of $267 billion, larger than the added defense expenses of the next ten powers.[3] Another primacy tactic is to seize control of critical geopolitical areas or "chokepoints" that allow a reduced number of defenders to foil a larger attacker; or that, conversely, confer the advantage to the attacker by removing the natural obstacles in its way that defenders may use for protection. For instance, the control of specific straits and capes conferred Great Britain a decisive advantage at sea over the other great powers by allowing it to control battlefleets crossing from one ocean or sea to another. As British Admiral John Fisher put it: "Five keys lock up the world! Singapore. The Cape. Alexandria. Gibraltar. Dover. These five keys belong to England, and the five great fleets of England will hold these keys."[4] Finally, the dominant state may also seek to expand its sphere of influence by bringing in additional regions, countries, colonies, or provinces. Typically, the targets of such expansion provide an advantage in a future hypothetical great power competition, by being located in the vicinity of states that may emerge as rivals, or due to their ownership of critical raw materials, energy sources, industrial and technological potential, or a large population. The dominant state may resort to various tactics for enlarging its sphere of influence such as conquest, purchase, coercion, or, as has been the case in the last quarter century, through the conclusion of alliance treaties and security partnerships, the granting of basing rights, or/and regime-change.

Subscribing to the status quo is compatible with moderate expansion, as long as the overall distribution of power and status remains roughly the same. Therefore, the acquisition of yet another piece of territory or the conclusion of one additional security partnership for the number one does not modify its position at the top of the hierarchy. This means that despite frequent accusations of representing world domination, primacy is not the same thing as the pursuit of hegemony. Instead, primacy is about holding on to the number one spot for as long as possible by making it harder for other great powers to contest it. That is to say that a dominant state deters opponents by creating situations of preponderance where challenges to its position become self-defeating propositions so that there is no incentive to attempt them. As William Wohlforth observes of the US in the post-Cold War, "any effort to compete directly with the United States is futile, so no one tries." By contrast, a would-be hegemon would seek to significantly alter the status quo in its favor to the point where any conceivable competition would be not just deterred, but completely annihilated. This makes primacy a fundamentally conservative endeavor aimed at consolidating and sharpening existing hierarchies, not at upending them.

The most serious drawback of primacy consists in its proneness to generate security dilemmas. Practitioners of primacy are genuinely convinced that their state's preponderance is a benevolent one, whose consolidation ends up benefitting everyone in the system. Consequently, they believe that the preponderance of their state should be welcomed and endorsed by any other peace-loving states as preferable to that exercised by any other great power; and that, furthermore, any opposition is either unwarranted or proof of nefarious intentions. Yet, in so doing, they are oblivious to the fact that their conservative and benevolent intentions are not self-evident. As Herbert Butterfield commented,

> you yourself may vividly feel the terrible fear that you have of the other party but you cannot enter into the other man's counter-fear, or even understand why he should be particularly nervous. For you know that you yourself mean

him no harm, and that you want nothing from him safe guarantees of your own safety.

Thus, a frequent result of status quo plus is that other great powers are alarmed by the expansion of the dominant state in their neighborhood and respond with counter-moves to what they perceive as aggression.[5] Hence, it is imperative for the dominant state to accompany primacy with judicious tactics of driving wedges between the other great powers to "reduce the number and strength of enemies organized against it." Primacy works best when the dominant state can count on the support or the neutrality of at least some of the great powers and of the regional powers. Wedges can be driven through selective accommodation by offering pay-offs to some parties (bribes, concessions in territory, security, status, or markets), while withholding them from others, or even punishing them. Another wedge tactic is confrontation or exerting pressure against a particular "weak link" coali-tion member, forcing it to seek out support from the other members. However, if these would be unable or unwilling to come to its rescue, the target would then be forced to come to an understanding with the dominant state.[6]

A major complicating obstacle in using wedge tactics is that dominant states favor unilateralism. Rather than including allies in their decision-making through consultations and concessions, dominant states lay down their preferred course of action and expect allies to fall into line. If, however, allies prove unsupportive, lukewarm, procrastinating, or exploitative, the number one simply moves on its own, considering it has sufficient power to achieve its objectives without pay-ing attention to other actors' objections. But while this manner of proceeding maximizes effectiveness, it also alienates would-be sympathizers, depriving the dominant state of their support or neutrality.

Primacy is not an option for every status quo great power. It is predicated upon capitalizing on one's *sizable* advantage over the other great powers to ensure a continual lead. This is impossible to do without enjoying such advantage in the first place, which is why primacy is restricted to the number one. However, not all dominant states necessarily endorse primacy. Doing so depends on two condi-tions. The first is if the dominant state's supremacy rests on tenuous geopolitical foundations: either it is open to a possible attack in the future by its neighbors, or its dominance rests on the control of areas or regions that may themselves be vulnerable to attack. If this is the case, there is likely to be an increased premium in consolidating one's holdings to ward off future dangers. The second condition has to do with whether the dominant state declines relative to its competition or maintains itself as an unrivaled number one. If decline has set in, with the dominant state being overtaken or caught up in multiple dimensions of power, primacy is no longer relevant, because this is no longer a matter of conserving a comprehensive advantage that has already vanished or is on the edge of vanish-ing. Instead, the declining number one would be interested in either using contain-ment to keep ahead of the competition, or a declining power strategy to recoup its losses or limit further ones. As such, primacy is a grand strategy for a number one at its apogee.

Concert

Concert designates a great power's grand strategy of promoting mutual conces-sions and cooperation between the members of the great power club with the aim of keeping in place the existing distribution of power and status among them. In so doing, the promoting great power ties down its competitors by determining them to submit to common norms. While the great power following a grand strat-egy of concert is bound as well by these rules, it has the least ambitions to satisfy, and, as a result, stands to gain the most and lose the least through the preservation of the status quo. This grand strategy is best seen in the case of Austria in between the Congress of Vienna (1815) and the Crimean War (1854), and, arguably, con-temporary Germany and Japan.

The term concert was originally used in the eighteenth century to refer to an ad-hoc temporary agreement to cooperate between two or several great powers. However, by the time of the Vienna Congress, it designated an informal long-term association of great powers consulting and, on occasion, acting together.[7] A concert grand strategy thus involves the promotion and conservation of such an association. It is useful to think of concert as a multilateral version of the tac-tic of engagement, understood as "the attempt to influence the political behav-ior of a target state through comprehensive establishment and enhancement of contacts across multiple issue-areas." In traditional state-to-state engagement, a state promises another to set up profitable political, economic, diplomatic, military, or/and cultural contacts in return for good behavior. If good (essen-tially cooperative) behavior is not forthcoming, it can cease or scale down these contacts, or in other words, disengage. The concert strategy is simply taking engagement one step further, in the sense that great powers engage with each other, by providing concessions to each other on their vital interests, and by pledging continued future cooperation in return for reciprocity. A great power pursuing a concert grand strategy does not buy off challengers with unrecipro-cated concessions as in appeasement, but instead changes their mind by affecting their cost-benefit calculations through socialization.[8] Accordingly, the concert grand strategy has a transformative role, just as traditional engagement aims to promote democratization and liberalization, turning actual and potential rivals into partners.

The key advantage of pursuing a grand strategy of concert is that it maintains the status quo at a bargain price. This is particularly important for established powers since they do not command the large resources available to the number one. The latter may find it more profitable to either embrace primacy, and con-solidate its already privileged position, or to resort to containment, answering challenges through the threat or the actual use of force from a position of superior military strength. But, for an established power of limited means, going to war or being on a perpetual war footing on behalf of the status quo may ruin it. It is therefore more beneficial from its point of view to ensure that challenges are never issued in the first place. This requires solving the two issues that motivate challengers to contest the status quo: reconciling the losers or have-nots from the status quo to the existing order, and at the same time convincing the winners not to press on their existing advantage.

The grand strategy of concert proposes to do this by determining the rest of the great powers to abide by three cardinal rules. The first rule is that each great power guarantees the others' territory, security, sphere of influence, and status for as long as they refrain from endangering the status quo. As Friederich Gentz, the Secretary of Austrian Foreign Minister Clemens Metternich, and the person who, as the Secretary of the Congress at Vienna, wrote up most of its documents, remarked:

> Europe seems to form but a great political family, reunited under the auspices of an areopagus [the five greats powers] of its own making, whose members guarantee to each other and to each of the interested parties, the quiet enjoyment of their respective rights.[9]

This is not an optimal arrangement from the point of view of any individual great power's ambitions, but it may work as long as each finds its terms tolerable. As mentioned in Chapter Two, according to prospect theory, status quo states prefer smaller but certain gains to larger but uncertain ones; at the same time they go to great lengths to reduce the possibility of loss. Accordingly, a general guarantee from their peers insulates them against loss, even if they have to provide a reciprocal guarantee.[10] This guarantee is attractive particularly when great powers are made aware of the magnitude or certainty of losses in case of conflict, typically in the aftermath of particularly destructive systemic wars, such as those of Napoleon's. The second rule is that, although differentiated by their power and status, each great power enjoys the same rights, and as a consequence, it is extended similar deferential treatment by its peers. The least great power commands, at least formally, equal privileges and immunities with the number one, punching above its real weight. At the same time, no member of the great power club is in danger of being humiliated. Since its high status is never questioned, it does not have to resort to force to defend it.[11] The third rule is that great powers show solidarity (hence the term concert) on important international questions, especially in crises, in which questions of using force internationally arise.[12] The reason is to avoid the risk of a general war caused by the counter-intervention of another club member. This outcome is avoidable if the would-be intervener demands the sanction of the other great powers beforehand through consultation. In this manner, the great powers that oppose intervention are able to make their objections known beforehand, so either they can be properly compensated, or, if this is impossible, the projected intervention is abandoned. As seen in Chapter Two, this rule of avoiding trampling on other great powers' interests has become over time so internalized that the very role-identity of a great power has become associated with recognition, i.e., being consulted by its peers at critical junctures.

The net effect of accepting these three rules is not just that great powers cooperate with each other, but more importantly, that in so doing, they give up on attempts at changing the existing distribution of power and status. This works to the great benefit of the established great power promoting the concert grand strategy, since it has no expansionist agenda to begin with; because it does not have to spend any military or economic resources in obtaining this result; and because it also gets in the bargain equal perks with the strongest powers, including the right of veto on questions touching upon its vital interests.

The greatest limitation of concert lies in that the great power pursuing it can rely only on persuasion and manipulation in order to get the other powers to play by concert rules. Persuasion involves convincing the other great powers that they have common interests, best served through a concert, and offering real evidence of advantages in the form of concessions. This, however, may prove prohibitively complex as some of these concessions come at the expense of other powers, who in turn have to be offered compensation. Meanwhile, manipulation plays on the other great powers' decision-makers' convictions, stereotypes, prejudices, and psychological proclivities in order to determine them to favor concert. However, manipulation may fail if one has to deal with different leaders in charge, who may bring to the table a different outlook on international relations. Thus, concerts tend to decay over time in the face of political change, and as the costs of the last systemic war fade from memory.[13] The crucial point is that a great power pursuing a grand strategy of concert is not allowed to use coercion in order to get other great powers to fall into line. Coercion not only violates the rules of concert, but it also makes it impossible to be achieved for an established power which has to take on far mightier adversaries. To this extent, if other great powers are unresponsive or resistant to its diplomacy, there is little that the concert manager can do about it.

Primacy: France and the status quo

From the second half of the seventeenth century and until the first decade of the eighteenth century, France outperformed the competition in nearly every status dimension. It had the largest population in Europe at 19 million people, three times as large as Spain's and England's, and double that of the Holy Roman Empire. Under Louis XIV, France fielded the largest and best organized army on the continent of 200,000 men in peacetime and doubling that number in war. As a result, France was able to undertake major military ground operations simultaneously on no fewer than four fronts. It also deployed the largest navy surpassing the traditional sea powers England and Holland, and, although later falling behind these two powers' combined forces, it remained the best navy in qualitative terms. Economically, it was the most productive country in the system with an estimated GDP of around $19 million in 1700, compared to Great Britain's 10 million, Holland's 4 million, and Austria's 2 million. True, Holland and England were comparatively richer on a per capita level and were also superior in trade and industry, but due to the reforms and innovations of Jean-Baptiste Colbert, Louis XIV's finance and navy minister, France had emerged as a formidable competitor to both. It counted as well on "the finest diplomatic corps in Europe," being represented in every major European state by an ambassador and in a multitude of the smaller ones by residents. It was the only state to have a centralized ministry of foreign affairs complete with clerks, translators, archives, and diplomatic training service. Furthermore, by 1715, France was the owner of the largest colonial empire in North America, the Caribbean, West Africa, and India, dwarfing both the British and the Dutch possessions. French predominance had become so taken for granted that the court of Versailles set for the next century the standard to follow in administrative, military, and diplomatic organization, and also in terms of laws, etiquette, arts, literature, architecture, and fashion. Also, by 1715, French had come to supplant Latin as the international common language, with treaties between states being written down in French.[14]

Yet, French supremacy coexisted with French vulnerability. Imagine the following picture. In the south, France was bordering Spain. But in the east, France also had Spain as a neighbor: in the north-east, in the Artois (the region around Arras and Lens), French Flanders (the region comprising Lille and Dunkirk), and the Spanish Netherlands (current Belgium); and in the south-east (in Milan and the Franche-Comté—the region around Besançon). And in the center-east, France neighbored Alsace and the duchy of Lorraine, areas it contested with the Holy Roman Empire, headed by Austria. As, at the time, the Habsburg dynasty ruled in both Madrid and Vienna, France was effectively sandwiched between enemies. Its position was all the more precarious as Paris was dangerously close, a mere 90 miles, from the Spanish border in the east. In 1636, during the Thirty Years' War, the Austrians advanced with little resistance to the outskirts of the capital, prompting the French to seriously consider evacuation. Breaking out of this encirclement was correspondingly priority zero for French decision-makers such as the French Prime Minister Armand du Plessis, best known as the Cardinal de Richelieu (1625–1642).

Perhaps the first genuine practitioner of raison d'état, the intellectual ancestor of grand strategy, Richelieu avowed to "perpetually arrest the course of Spain's progress," even by taking the scandalous step at the time of concluding alliances with "heretic" Protestant powers (Holland and Sweden) against fellow Catholics. As he claimed, France was the natural counter-weight (*contre-poids*) to Spain. But Richelieu's goals went far beyond balancing. As he explained to the French king, Louis XIII, the ultimate objective was to make him "the most powerful monarch and the most esteemed prince in the world," in other words, to elevate France as number one at Spain's expense. Only under the benevolent protection provided by a dominant France, could other states truly enjoy their liberty, since the best interests of France and of the smaller states of Europe went hand in hand. The most interesting part of Richelieu's plan consisted in how he envisaged to transform and maintain France as the dominant power. He argued that France should "seek only to fortify itself and build and open gateways for the purpose to enter all the neighboring states to guarantee their protection from Spanish oppression." To this end, he advised acquiring a number of strategic key points—gateways (*portes*)—into Germany, Switzerland, and Italy. These gateways were supposed, on the one hand, to close French territory to penetration from outside, but, at the same time, on the other hand, to open the territory of its neighbors to French offensives, under the ostensive motivation to shelter them from Spain. What Richelieu left unsaid was that if the danger from Spain were to ever recede, through the control of the gateways, France would have been able to carry out intervention at will, while being quasi-immune from attack itself.[15] And therein lay the whole problem: if the eastern border of France had been controlled by any other great power, France would have been in constant danger. But if it had been France that owned that territory, then every one of its neighbors would have had to put up with France as a potential overlord. Thus, the control of areas such as Belgium, French Flanders, Alsace, and Lorraine remained perhaps the most insoluble security problem in the world from the days of Richelieu up to 1945. Effectively, the battles in both world wars were fought in almost the same locations as the battles of Louis XIV.

By 1659, France had made commensurable strides in attaining Richelieu's target, which all but guaranteed that its dominance would be unassailable. At the end of an exhausting war of nearly a quarter of century against Spain, France, led by Richelieu's hand-picked successor, Cardinal Mazarin, annexed territory in its south (the Roussillon) that made the Pyrenees mountain range the effective border between the two states, as well as most of the Artois. These were not impressive gains in extent, but they marked the beginning of the decline of Spain, exhausted economically, stagnating demographically, and on the verge of a major dynastic crisis. Since this crisis would prove crucially important later on, it may help to sum up the main points. The Spanish king had two children: a daughter from his first marriage, the Infanta Maria Theresa, and a sickly son, Charles II, as his heir. In the eventuality of the latter's death, the question of how to distribute Spain's extensive possessions would have arisen. At the time, Spain controlled, besides Spanish territory, Latin America (minus Brazil), the Philippines, present-day Belgium, swaths of present-day France, and most of Italy. Hence, the cunning Mazarin sought to secure French ownership of at least some of Spain's European imperium by arranging the marriage of Maria Theresa to the French King Louis XIV. It is true that, in the marriage contract, Maria Theresa was giving up all her pretensions to the Spanish inheritance, but on condition of Spain paying France an exorbitant dowry of half a million crowns, which, predictably, was never delivered. Therefore, conditions were ripe for Louis XIV to later on press his claim to select parts of the Spanish inheritance.[16]

Louis XIV took advantage of these propitious circumstances to secure France's position as number one. As one scholar put it, his grand strategy essentially consisted in increasing "the grandeur of his State and of his House, so that his own pre-eminence as the greatest king in Christendom would be beyond dispute."[17] This was to be done by putting into practice Richelieu's gateway vision by acquiring and then fortifying a series of defensive positions on France's east. Louis could count on the commensurable strategic authority of Sébastien LePrestre de Vauban, a military engineer from the small gentry. Vauban had made his reputation in the successful pursuit of 42 sieges and in the building of 150 fortresses. Eventually elevated to maréchal of France, he argued that France was placed

> in the middle of the most considerable powers of Christianity, equally vulnerable from attack from Spain, Italy, Germany, England, and the Netherlands.

... France had attained today a high degree of elevation, which renders it formidable to its neighbors, so they are all interested in its ruin or at least in the reduction of its power.

As such, France's current borders were indefensible, being intermingled with enemy possessions. In a war in which it suffered attacks on multiple fronts, Vauban reasoned, France could not have brought its entire strength against a single enemy, but instead had to face them with inadequate numbers on each separate front. Consequently, France would not have been able to hold on to any territorial gains it had made, because it would not have had enough troops to protect them. It might have been able to defeat its enemies piece-meal, but any reversal of fortunes would have spelled disaster. Besides, the costs of keeping armies on every side would have been ruinous. Hence, the solution Vauban advocated was the so-called *pré carré* (literally meadow field, but with the meaning of field of duel)—a double or even triple barrier of fortifications in key positions on enlarged French borders, which would have made France literally unassailable. The *pré carré* meant both giving up forward places that could not have been protected properly, and securing new pieces of territory to form continual fortified lines, which could be defended efficiently by fewer men. Like Richelieu's gateways, Vauban's *pré carré* had an aggressive defense ("la defense aggressive") role: being secure itself by controlling chokepoints, France could strike at will against its vulnerable opponents. However, this was not a call for upsetting the status quo, in other words, for hegemony. Vauban did not argue for territorial annexation without end or wherever annexation was feasible, but rather for redrawing France's frontiers in a manner that would have made foreign attack futile. As the inscription on a monument honoring Vauban which Louis XIV erected on the Rhine stated, the goal was to guarantee the perpetual security of France ("Securitati Perpetuae").[18]

Essentially, this was a patient, painstaking, piece-meal strategy of status quo plus, not a call for unlimited aggression. As a result, Louis' acquisitions were not entire provinces, but rather consisted of cities and strips of territory that could be promptly fortified and turned into strategic strongholds.

Louis' first move to bring about this agenda was made against the Spanish Netherlands in 1667. Following the death of his father-in-law, Louis argued that he was entitled to a part of the Spanish inheritance, on account of having never received the promised dowry for his wife. The Spanish Netherlands were an obvious target both for strategic reasons, but also because Louis invoked as a legal point the so-called right of devolution, which figured in the law statutes of the targeted province, and which specified that the eldest child from the first marriage could not be excluded from inheriting in favor of a child from the second marriage. The French king did not intend to claim the whole of the province, but only the parts needed for the defense of France, announcing his willingness to return the rest to Spain in return for the Franche Comté. Faced with French forces more than twice their size, the Spanish suffered utter defeat. But at the very moment when France seemed on the verge of realizing its long-standing objective of securing the north-east, it ran into resistance from a quarter it had not anticipated: its own ally, Holland. The Dutch were said to be following the adagio "gallus amicus sed non vicinus" ("the French make good friends, but not good neighbors"). It was one thing for Holland to be bordering a weak declining power such as Spain, but quite another to share a frontier with a dominant France. The problem was that no matter how many insurances of good behavior France would have offered the Dutch, they could not trust France's good intentions long-term. Accordingly, in 1667, Holland stroke a hasty anti-French alliance with its former enemy England, and brought along Sweden. Faced with a hostile declaration by this triple alliance, Louis backed off, but not without securing a number of gains. As negotiations went on, he expanded military operations to the Franche Comté, and then used it as a bargaining chip to obtain the rest of the Artois and parts of French Flanders. Nonetheless, for all the benefits, the war of devolution did not solve France's long-term encirclement dilemma, and Louis placed the blame squarely on Holland.[19]

Louis now realized that in order to settle the frontier against Spain in the north-east and the south-east, he had to first dispose of the Dutch. As his defense minister, the marquis of Louvois, put it, "the only way to conquer the Spanish Netherlands successfully is to humble the Dutch, and, if possible, to destroy them."[20] Louis XIV prepared his attack very carefully by driving wedges between Holland and its two alliance partners. At the time, it was a common custom throughout Europe for state officials to receive pensions and gratifications from foreign powers, as long as they stayed nominally loyal to their sovereign. Some lesser princes, espe-cially in Germany, even welcomed the practice, as it spared them the expense of paying these officials out of their own pocket. No state was more generous in this respect than France, with the full expectation in Paris that this money, effectively a bribe, would result in lobbying in favor of French interests. Thus, at one point or another, England, Savoy, Sweden and a large number of German states were on France's payroll. This is how Louis got England to pledge to take up arms against Holland in return for a yearly subsidy of 2 million crowns. Similarly motivated, Sweden then exited the alliance and was brought to promise to provide support if any German prince intervened to help Holland. Not that Louis had failed to take provisions on this side as well, the electors of Cologne and Munster having promised in return for generous French payments to allow passage to the French troops and even join the hostilities on the French side. Finally, Austria, with an eye on Hungary, threatened by the Ottoman Empire, which was in turn egged on by Louis, was induced to sign a neutrality agreement.[21] Accordingly, when Louis declared war in 1672, the Dutch found themselves fully isolated, with only weak Spain as an ally.

As one historian put it, Holland collapsed like a bad soufflé. Facing full occu-pation, the Dutch opened their dykes, flooding the country and halting the French advance. This expedient provided the respite new Dutch stadtholder William of Orange needed, s in order to gather allies. Spain was interested, knowing that after Holland, its possessions would be next; and by now the alarmed Austrian emperor and German princes were also prepared to help. Meanwhile, England failed to defeat Holland at sea and exited the war. Faced with a formidable combination of enemies, Louis once again chose to give ground. Rather than holding to his new conquests, he pulled back and engaged in a war of attrition, while playing on the allies' dissensions. This tactic worked out very well. The German side of the coalition believed that the worst had passed; meanwhile William was bedeviled by domestic calls for peace. Consequently, the coalition was weakened, allowing Louis to strike against the Spanish. Thus, even though France had to agree to a restoration of the territorial situation before the war with Holland at the Peace of Nijmegen of 1678, it still wrestled away from Spain and the Holy Roman Empire the whole of Franche Comté, the Lorraine, parts of Alsace, and 11 strategic towns in French Flanders. In so doing, Louis had made serious strides into achieving the *pré carré*: one observer put it, France had become the arbiter of Europe: no state could find either security or profit without the friendship of the French King.[22]

There were, however, gaps still to be filled out in the French rationalized line of defense, especially on the German side of the border. To draw away Austria, Louis then resorted to encouraging the Ottoman Empire to go on a major offensive in 1683. While the German world had its hands tied fighting the Ottomans, Louis unearthed an old statute that mentioned that any territory that, at one time, had been formerly acquired by the Crown of France could not have been at any time given away. Armed with this legal argument, France created several courts under the label of chambers of reunion in Alsace, the Franche Comté, and Lorraine, which promptly ruled that France had rights on additional territory which it had once owned, together with its dependencies. But no one actually had bothered to specify where these dependencies were supposedly located. As a result, France's border crept further east, by claiming parts of the Spanish Luxemburg, as well as the whole of Alsace. Then, on the same day, September 30, 1681, French troops, dispensing with further legal niceties, took control of Strasbourg, the gateway into Germany, on which France had no claim, and also occupied the fortress of Casale, which guarded the access from and to Italy, and which Louis had bought from the Duke of Mantua. A desperate, belated effort by Spain to save what remained of Luxemburg ended with the loss of the entire province—France did not want it for its defensive value, but intended to use it as a bargaining chip in any future negotiations. Vauban soon had fortifications built providing France with protection from Flanders to the Alps, and Louis struck a medal announcing that France had closed the gateway to Germany ("Clausa Germanie Gallia").[23] The *pré carré* had been made into a reality. This was the moment when France enjoyed maximum power. The problem was that the situation from that point onward deteriorated.

The Ottoman Empire suffered catastrophic defeat and severe territorial losses to Austria in Hungary and Transylvania, sending the prestige of the Austrian emperor Leopold I to an all-time high. An even worse development occurred in England. All through his campaign to consolidate France's position on the continent, Louis had relied on the benevolent neutrality, if not outright support, of the Stuart kings Charles II and James II. But when the latter sought to publicly embrace Catholicism, he was faced with a domestic explosion which caused his replacement with his son-in-law and Louis' sworn enemy, William of Orange. Louis offered initially his support to James II, but being rebuffed, took consolation in the thought that the upcoming civil war would last a long time keeping both the Dutch and the English busy. It proved one of the worst mistakes, if not the worst, that Louis had ever made, as William became the ruler in both England and Holland, bringing both countries' combined resources to bear against France. Louis made a further error by giving up on the wedge tactics which he had employed beforehand to such good effect. Judging that the German princes were too fickle and too demanding, he stopped trying to woo them. One consideration behind this measure was that France was by now powerful enough on its own not to require allies; and another that the "shower of gold" of French pensions and gratifications had come to a halt due to the increased expenses of building up Versailles and enlarging the French army.[24] But this was precisely the wrong time to change tactics. France was advancing in the Holy Roman Empire, and, consequently, its advance was as alarming for German princes as it had been for the Dutch. With no select incentives to offer, most of them rushed into the welcoming arms of Austria. This meant effectively that Louis had unwittingly contributed to the creation of a grand anti-French coalition consisting of Austria, the Anglo-Dutch combination, Spain, and most of the German world.[25]

Louis knew that the moment the Ottoman Empire completely collapsed, the coalition would turn west against him. So, he sought to fight on terms favorable to him, by denying the use of valuable territory to the coalition. He thus launched a scorched earth expedition in Germany, against the Palatinate, burning to the ground venerable cities such as Worms and Mannheim so that no house was left standing, in order to prevent the building of future fortifications. The sack of the Palatinate proved the last straw for the German princes, who launched war against France, soon joined by nearly everyone else. In this war known as the War of the League of Augsburg (1688–1697), a dominant France was facing alone almost the entirety of Europe, whose avowed goal was to push it back to the territorial limits it had occupied in 1659. It was on that occasion that Vauban's *pré carré* showed its usefulness. After nine years of futile attempts to break through the barrier of French fortifications, the allies had to settle for a draw. To his credit, Louis changed tack during the war by fomenting and exploiting wedges in the coalition, so that by the end of the fighting, Savoy had changed sides in exchange for Casale, which however the French razed so it could not be used against them; and William of Orange agreed to make peace in return for Louis acknowledging him as King of England. In the final peace treaty signed at Ryswick, France made only concessions in the Lorraine and in Luxemburg. But Louis kept several fortresses in Lorraine as well as Alsace and Strasbourg.[26]

So far, Louis' grand strategy had been a considerable success, but it placed France in a particularly dangerous spot, because due to the security dilemma, it was viewed with suspicion and hostility in nearly all European quarters. This meant that it should have judged itself content with what it had obtained and refrained from seeking any other gains, particularly considering the cumulative economic toll of nearly uninterrupted war. The problem was that, despite realizing that this was a bad move, Louis could not escape another general war that came close to undoing everything that he and his predecessors had worked for.

The war in question involved the above-mentioned Spanish succession. Since the sickly Charles II had been unable to beget any children, this meant that the whole Spanish Empire could go either to an Austrian Habsburg or to a French Bourbon. It went without saying that any effort to join either Austria and Spain or France and Spain under one ruler would have been rejected by everyone else, because they would have de facto created a hegemon. Therefore, to avoid another general war, Louis negotiated with his nemesis William of Orange a partition treaty of Spanish possessions between France and Austria. But there were several problems in there: the treaty did not include Austria, and Spain was not taken into account. From the point of view of Madrid, the situation stood thusly. As long as Spain kept all of its possessions, it remained a household great power, albeit past its prime. But if dismembered, it would have been hardly pressed to stay in the great power club (a verdict that proved eminently correct). Hence, the Spanish courtiers pressed their king to write a will naming a Bourbon, Louis' grandson Philip of Anjou the sole heir, on condition he would not have given up anything that belonged to Spain. If Philip declined the offer, the entire Spanish Empire would have gone to a Habsburg. That was a very shrewd offer, since the Spanish elites calculated that only France had the requisite power to keep the Spanish Empire in one piece.

Hence, Louis faced an unsolvable dilemma. If he rejected the Spanish offer, he would have confronted an Austrian compact on both the south and the east, with no guarantee of help from England and Holland. If, however, Philip accepted becoming King of Spain, but France repudiated him by sticking to the partition treaty, it might have meant that Louis would have had to make war on his own grandson in conjunction with Austria, receiving nothing for his efforts because Austria was not part to the partition treaty. Basically, France would have fought to aggrandize Austria. Both options were terrible, and the only alternative left was therefore war. At least, Louis reasoned, by taking over the entirety of the Spanish holdings, France would have removed for good the threat of encirclement by turning Spain (and hence the Spanish Netherlands as well) into a French ally, and would not have fought alone this time, for Spain would have joined it. It is doubtful that Louis believed he could unite France and Spain under one crown or that he could keep every bit of territory that Spain owned, but by fighting, he could make a solid bargain for at least some choice parts, particularly the Spanish Netherlands and Italy.[27]

Nevertheless, the unforeseen intervened: everything that could have gone wrong for France went horrifically wrong. On paper, France was stronger than it had been in the previous war, with the addition to its camp of Spain, Savoy, Portugal, and even a number of German states. But in practice, it meant that France had to move its troops beyond its fortifications to help out its new allies. French general-ship was at an all-time low, while Great Britain and Austria fielded the best gener-als of the age in the persons of the duke of Marlborough and of Prince Eugene of Savoy, who ironically had been denied a command by Louis in the French army. France suffered three crushing defeats at Blenheim, Ramillies, and Oudenarde, losing a combined number of 70,000 soldiers. The Spanish Netherlands and parts of French Flanders fell to the allies, Portugal and Savoy switched sides, and it seemed that Philip V, the king's grandson, was on the verge of losing the war in Spain. The *pré carré* delayed the invasion of France, because every single component of the two lines of fortresses had to be conquered through a prolonged siege, but eventually, both lines fell by 1710, opening the door to Paris. On top of these calamities came the disastrous great winter of 1709, with temperatures in France falling twice below −20 degrees Celsius, 20,000 dying from the cold, and crops failing. For the first time in his reign, Louis found himself forced to sue for peace. He offered extensive concessions: most of the Spanish Empire, save a minor part of Italy as some form of compensation in return for Philip giving up his claim to the Spanish crown. But smelling blood, the allies went further. They now demanded the surrender of most conquests since Louis' accession to the throne. It was proof of how desperate the situation of France had become that Louis actually agreed to "these unheard-of conditions." But the allies then made a mistake of their own: they sought to humiliate the haughty Sun King, by asking that not only Philip give up Spain, but also that in case he refused, which was likely, that Louis declare war on his own grandson. This meant not just a loss of capabilities and the deterioration of its security for France, it was also a tremendous loss of status, with the kingdom reduced to the condition of a client state, and Louis forced to do the allies' bidding. This the Sun King would not stoop to do. Instead, he did the unthinkable for an absolute monarch. He had started his reign by proclaiming that "l'état c'est moi" (I am the state). He ended it with the deathbed message that "I am leaving you, but the state remains forever." Louis appealed to the people, by writing a manifesto publicly read throughout France. In it he candidly explained that to obtain peace he would have had to accept "conditions highly dangerous for the safety of my frontier provinces", that every concession he had provided had only led to more demands, and that the demand to make war on his own grandson went against every humane sentiment. Louis concluded that

> even though my affections towards my people are no less great than those I feel for my own children … I am certain that they would be opposed to accepting [peace] upon conditions contrary to both the justice and honour of the French name.[28]

Louis' appeal galvanized France. Despite being near starvation, it fought on. Food supplies were brought in by ship. The French nobility and even the common people sacrificed to the war cause—new taxes were imposed and Louis melted the silverware in Versailles to keep financing the war. By that time, the expense of nine years of war had already taken a large toll on the allies, as well. All it took was for France to prove it was not going to collapse in the near future and that the war would go on without an end in sight. This proof was provided by the battle of Malplaquet, where even if the French were defeated, they imposed a heavy cost on the allies, with 21,000 killed and wounded against 11,000 French casualties. The French situation also improved in Spain. First, Philip V, who proved more popular than his counter-candidate with the Spanish, miraculously prevailed over the allies' forces. Then, the death of the Austrian emperor meant that his heir, the chief Habsburg contender to the Spanish throne, also inherited the position of Holy Roman emperor, so that any further ally endorsement for his claim on the Spanish inheritance would have resulted in the re-constitution of a Habsburg superpower. In this context, skillful French diplomacy won what could not have been won by arms—first, Savoy was induced to again leave the war. Then, it was the turn of Great Britain, where Marlborough's wife Sarah Churchill lost favor with Queen Anne, which led to the fall from power of her husband, to press for peace. Britain was followed by Holland, where commercial elites disgruntled by unceasing warfare complained about their trade profits. Soon, it was only Austria left facing France, and a succession of last-ditch French victories in Germany convinced Vienna to come to terms. The resulting settlement at Utrecht was a compromise largely effected at the expense of Spain, but sparing France itself. Partition was unavoidable: while Spain's possessions in Italy and Belgium went to Austria, the rest remained in the hands of the Bourbon king. Thus, Louis had established his dynasty in Spain and in the same move prevented an Austrian takeover there, the main goal of the war. As John Lynn puts it, the proclaimed goal of the allies was "no peace without Spain," but the peace obtained was certainly peace without Spain. Louis, moreover, had succeeded in doing so, while sacrificing little of the frontiers gained at Ryswick (essentially the return of Lorraine to its duke, which proved temporary, and giving concessions on the border with Savoy).[29]

Was France's grand strategy successful? If Louis' goal had been to obtain hegemony by unifying France and Spain, then he had indeed suffered a crushing defeat, as the prospect of unification was forever quashed. But France had never nurtured such ambitions.[30] Alternatively, seen from the point-of-view of the objective of conserving French primacy and giving France strong frontiers, Louis' grand strategy, although costly, was largely effective. Any threat had been removed from the south where now France shared a border with an ally. In the east and north-east, France could count on an iron barrier formed of Artois, Flanders, Alsace, and the Franche Comté to protect Paris. Strategically, France was far more secure in 1715 than it had been in the 1640s.[31] From this improved defensive position, France maintained itself as the number one power for another century, until the conclusion of the Napoleonic Wars, and as the chief continental power until the unification of Germany. The fact that France is currently the second largest state in Europe in terms of territorial size after Russia is largely due to Louis XIV's grand strategy.

Concert: Austria and the status quo

Austria in 1814 was no pushover. Victory over Napoleon had left Vienna the owner of commensurable estate in Central and Eastern Europe, comprising the territories of contemporary Austria, Czechoslovakia, Hungary, Croatia, and Slovenia, together with parts of Italy (Milan and Venice), Romania (Transylvania and Bukovina), and Poland and Ukraine (Cracow and Galicia). Austria was the second largest power in size and population at 37.5 million after Russia. Yet,

Austria suffered from an uneven capabilities portfolio. Most of the area under its control remained agrarian—its per capita levels of industrialization and the relative share of manufacturing output were basically the same in 1800 and three decades later, while every other great power's had increased. The war against Napoleon had bankrupted the state and had been fought on British subsidies that vanished after victory. To make matters worse, Austria had inherited from feudalism an old-fashioned administration, personally supervised by the emperor, that routinely rejected much-needed political and economic reforms. As such, Austria did not have, nor could obtain, the resources to develop and maintain a strong army: military expenditures were "a shield which weighed down the rider." Nominally, Austria's army stood around 230,000 soldiers—but, due to low morale and poor equipment and organization, it probably could field only a third of this number. As one decision-maker quipped, Austria was "armed for perpetual peace." As Friedrich Gentz concluded, in these circumstances where everything combined to bind Austria to a peaceful system, it was paramount for Vienna to maintain the status quo for as long as possible.[32]

However, Austria faced simultaneously a wide array of threats stemming from its geographical location. Should trouble have occurred in either Western Europe, Eastern Europe and the Balkans, or Germany, even if it was not directly involved, Austria would have been first in line to feel the impact. As Henry Kissinger put it, Austria was the seismograph of the continent. It is no coincidence that both world wars started precisely in connection to territories that had been part of the Austrian Empire: Bosnia-Herzegovina in World War I, and Poland in World War II. But apart from the threats from abroad, Austria also had to cope with the threat from within. Only slightly more than a third of its population were ethnic Germans. In the age of the French Revolution, which had opened the door to liberal and nationalist demands for political rights, including national self-determination, the Austrian Empire stood as an antiquated oppressive edifice, the so-called "prison of nations." This meant that any foreign complication could set off internal revolution, which could unravel the entire state.[33]

How did the Austrian Empire manage to not only survive for another century but to actually thrive up to the age of Bismarck? The credit should go to the grand strategy of concert put into place by the Austrian foreign minister, and later on Chancellor Clemens von Metternich. Metternich remains a polarizing figure to this day. Most of his contemporaries execrated him. One commentator wrote that "never was a man more feared or detested" and that liberty had never had as dangerous an enemy. The British foreign minister George Canning called him "the greatest rogue and liar on the Continent, perhaps in the civilized world." Napoleon agreed: he observed of Metternich that for someone to lie all the time was too much. Yet, more recently, Kissinger portrayed him as the quintessential virtuoso diplomat.[34] Indeed, Metternich may have been guilty of excessive arrogance and duplicity, yet he also managed through sheer finesse and dexterity to achieve more than Austria could have ever hoped to do on the battlefield.

Metternich's view of international relations was based on two considerations. First, if Austria could not afford to fight any great power war, it was essential to create conditions that would prevent any such war from starting in the first place. This was not just a matter of ensuring that Austria itself refrained from using military force, but also one of neutralizing at an early stage any inadvertent clash of the other great powers that could drag it in. Second, any liberal or nationalist revolution in Europe represented a mortal threat, because not only could the victory of the revolutionary principle in one context encourage its spread to Austria, but also because it could result in a revolutionary government which then would have proceeded, just as Napoleonic France had, to cause renewed war. It should be mentioned that with the exception of Great Britain, every one of the great powers had good reason to fear revolution, being led by absolutist governments, including France, where the Bourbons had just been restored (and even the British government was elected at the time by only 1% of the population). As such, anywhere, revolution was viewed as a deadly epidemic that endangered domestic and international order everywhere. This was not paranoia. From 1815 to 1848, revolutions occurred in Spain, Naples, Piedmont, Latin America, Greece, Belgium, France, and Poland. Therefore, it was crucial to check revolution from the start, if necessary, through foreign intervention.

Yet, it was equally clear to Metternich that Austria did not have the resources needed to act as the policeman of the continent. Among the great powers of 1815, Britain and Russia towered above the rest—it would have been impossible for Austria to stand up alone to either, and even a coalition of Austria, Prussia, and defeated France might have had trouble in dealing with an Anglo-Russian combination. Moreover, how could Austria have intervened militarily to repress revolution in other states without triggering war through the counter-intervention of the other great powers? The answer for Metternich was to turn both Austrian concerns into European problems, by convincing the rest of the great powers that these were issues that affected their vital interests as well. Accordingly, they had to be dealt with, not by Austria alone, but rather by a committee or federation of great powers. In this way, by subscribing to common rules of behavior, on the one hand, each committee member would have accepted self-restraint and at the same time helped restrain the other members; while, on the other hand, in case of revolution, the entire committee would have been called into action to extinguish the fire before it could spread. But the committee rules were formulated and administered by Austria. As Metternich ironically described his role, he was at the same time the coach driver of the European carriage and its doctor of revolutions.[35]

The trick was how to persuade the other great powers to adopt the worldview favored by Austria—in the form of a quid pro quo between a guarantee of their territory, security, and status, and their endorsement of the status quo. This problem manifested itself in the complex negotiations that took place for more than half a year from October 1814 to June 1815 in Vienna, which had as an object what to do with the Napoleonic empire and its satellites. To quote Gentz, while the representatives at Vienna spent their time enouncing "high sounding phrases [such as] 'lasting peace based on a just division of power,' … the true aim of the congress consisted, however, in the division among the victors of the spoils." This order of business proved particularly thorny for Austria, because war between the former allies could not have been ruled out either in the present or the future if this distribution failed to be at least tolerable; because in some cases several powers were laying competing claims to the same territory; and because in other cases, the expansion of a given great power would have left others at a disadvantage. For a compromise to occur, Austria and Britain made the ostensible case that every great power had to accept some sacrifice of its ambitions. Out of the five great powers, France, as a defeated party, was not to receive anything, but was to conserve everything it had owned prior to 1792, thus losing its empire, but not its own territory. This was not in any way punishing—if France had been weakened too much, Austria might not have been able to count on it to resist Prussia or Russia.[36] Meanwhile, Great Britain could be placated by being given a free hand outside of Europe, where it inherited colonial possessions from the Netherlands and France. In return, Britain made no demands in Europe. As for Austria, it was willing to generously give up its ownership of Belgium (the former Austrian Netherlands) and to accept a German Federation to replace the ancient Holy Roman Empire, which had been dissolved by Napoleon.

Only these sacrifices were no sacrifices at all from the point of view of Metternich. Knowing the ambition of France since the days of Louis XIV to control Belgium, he thought it wiser to transfer the troublesome territory to the Netherlands, in the process avoiding the risks of sharing a border with France. As for Germany, he would not have pressed the case for any German national unification formula anyway, for the simple reason that it would have delegitimized Austrian rule over other nations in much of the empire. Better, therefore, to create and preserve a patchwork German federation made up of weaker German states, which could be used as a shield against either France or Prussia. It is telling that Metternich, nevertheless, kept the presidency of the German Federation in Austrian hands, thus conserving the decisive influence in the area. In return for these "sacrifices," Austria retrieved the other areas it had lost to Napoleon, namely Milan, Croatia, and Slovenia, with the addition of Venice and the imposition of Habsburg rulers in Central Italy. Eventually, even though the Bourbons were restored in South Italy, they too depended on the support of Austria, so the entire peninsula was either directly or indirectly under Austrian control.

Here an ordinary decision-maker might have stopped, having achieved all he or she had set up to do. But Metternich remained worried concerning the pretensions of the remaining two powers: Prussia and Russia. Prussia was Austria's rival in Germany, and now laying claim to either the entire kingdom of Saxony, or to its former share of Poland. Austria could not concede Saxony to Prussia; yet at the same time, Russia claimed the whole of Poland, which was under occupation by half a million Russian troops, thus making impossible any Prussian gains in that quarter. To complicate matters further, Austria was also reluctant to see a Russia in full control of Poland, not just the part it had owned before the Napoleonic wars. Finally, Metternich was aware that if hostility flared up between allies, which was not fanciful since Prussia was threatening war, it risked undermining the rest of his carefully laid-out settlement and igniting the war that Austria was so dreading.

What followed was an exercise in masterful manipulation. Metternich resorted to seeming inactivity, a favorite tactic of his, by dragging on negotiations, first by feigning illness, and then by organizing a seemingly endless succession of balls, operas, music concerts, and hunts, to which he acted as the master of ceremonies. The Congress of Vienna soon became known as the dancing congress, on account of its excess of social occasions—only in the month of January there were more than 20 balls. The French foreign minister was not fooled by Metternich spending an estimated three fourths of his time on festivities, commenting that his "greatest art is to make us waste our time, for he thinks he is gaining by it." This apparent passivity was designed to force the other powers to show their cards first, allowing Austria to pronounce itself last, and benefit therefore from a maximum freedom of action, being the one actor with the most options left. Thus, instead of openly confronting Prussia over Saxony, Metternich announced that he was endorsing the Prussian claim, as a further sacrifice Vienna was willing to make in order to pre-serve the unity of the great powers. In return, he asked for a Prussian promise to join with Britain and Austria in checking Russia over Poland. What was going on was that, in so doing, Metternich was deceiving Berlin that Austria was in its corner while using it against Russia, and simultaneously ensuring that its effort to obtain Saxony would come to naught due to the opposition of the other powers, which he simultaneously orchestrated. The very endeavor to support Prussia was an elabo-rate trap. What Metternich left unsaid was that if Russia backed down on Poland, Prussia could no longer hold itself any pretensions to Saxony since it had already received its compensation: it could not have both. But if Russia did not back down, Austria could charge that the initial agreement was no longer valid, and withdraw its support for Prussia's claim over Saxony. Eventually, this proved to be the case, and by then Metternich could also count on support from a different quarter.

At the start of the Congress, in virtue of the many participants from all cor-ners of Europe—a total of 5 heads of reigning dynasties, 216 princely families, and 2,000 plus diplomats, not counting their retinues—it was established that, as a matter of practicality, all key decisions would be taken by the representa-tives of the four Allies (Russia, Great Britain, Austria, and Prussia). Yet, a further consideration was that France, although defeated and deprived of its satellites, was still a force to be reckoned with. As it sought to reinsert itself in the debate of key decisions, it attempted to have every participant included, counting on manipulating the majority votes of the small states of Germany and Italy. This was rejected: France was not to be trusted, and any deal with the former enemy would have been considered anathema.[37] This decision had two results. First, the great powers emerged as a European cabinet, federation, or concert, which took authoritative decisions for the rest of the states. Even though no one had elected them for the job, they took upon themselves the joint management of the system to the exclusion of other states, who tacitly submitted to their authority. Second, France, realizing which way the events were moving, changed tack by seeking above all to secure its presence alongside the other four powers. While this might have initially seemed repulsive to the great powers, given the benefit of months of negotiations and the risk of the allies coming to blows, it gradually became not just conceivable, but also wise. By waiting, Metternich had allowed France to gradually re-join the other four powers. It was therefore a simple matter to make sure that in return for admission to the concert, France would have opposed the cession of Saxony to Prussia. When the French objection was finally unveiled and was also supported by Britain, Austria only had to rally to their side, by endorsing their argument that the concert supported the independence of all states, great and small. Legally speaking, the great powers had no business of disposing of territory controlled by the King of Saxony, so that it was up to him to cede anything to Prussia. Prussia ended up isolated, because Russia would not have fought to secure it Saxony, and had to give up its demands (it eventually was ceded part of the state as punishment for Saxony's support of Napoleon , but little of its population and none of its industrial centers). But if Prussia could not have Saxony, this meant that it could receive compensation only in Poland. Now it was Russia that was confronted by all the other four great powers over Poland. Although, it managed to retain the lion's share, a new partition was unavoidable, with part of Poland going to Prussia, and, even Austria receiving its share.[38]

The cherry on top of the cake for Austria was that every great power found this settlement acceptable, considering that the alternative was renewed fighting, even though with the exceptions of Britain, and Austria, everyone had to accept much less than they had initially sought to achieve. Furthermore, the settlement became very difficult to undo, because, for all its defects, it was at least workable. What could be devised to replace it without reopening a giant can of worms? Once the settlement was revised in one instance, it would have ushered in all manners of subsequent revendications, which could only have led to war. Accordingly, the conservation of the distribution of power agreed upon at Vienna and which greatly advantaged Austria, had been transformed by Metternich into a general European concern. Any challenger would have been confronted not just by Austria, but by every other great power that stood to lose out by the revision.

Getting the great powers to commit to keep in check revolution proved equally arduous. Metternich took advantage of the opportunity created inadvertently by the Tsar of Russia Alexander I, who proposed the conclusion of a general alliance with the ill-defined purpose of reforming international politics in accordance with Christian morality. Metternich did not think much of the idea at first, dismissing it privately as a "loud sounding nothing," even though he, like nearly everyone else, signed on the Tsar's pet project that was to become known as the Holy Alliance. However, by 1820, a military putsch occurred in Spain and a revolution erupted at Naples, making Metternich reconsider the Holy Alliance's value. In the case of Spain, Metternich had originally preferred a "magisterial inaction" attitude, justified by two reasons: the most likely power to intervene against the revolutionaries would have been France, and in so doing Paris would have reasserted itself in a traditional sphere of influence; and Britain, on whose continued cooperation Austria depended to uphold the Vienna settlement, strongly objected to any prospect of French intervention. Therefore, for the moment, Metternich hid behind the British non-intervention argument that stated that the Congress of Vienna had not specified anything about interfering with the internal arrangements of independent states. But once revolution spread to Naples under the control of an Austrian ally, this became quite a different story. Hence the issue became: how to obtain a European authorization for Austria to put down the revolution in Naples, while at the same time, continuing to deny France's similar request about Spain? By now, both Russia and France were complaining about double standards, while in fact both were attempting to nibble away at the limits imposed on them by the concert. Thus, the Troppau Congress of October–December 1820 started, in which effectively Metternich managed to square the circle.

Once again, instead of rushing to repress the Naples revolution, Metternich preferred to wait and consult with the other members of the concert. This nonchalant attitude worked very well, because as the international situation was getting tenser, the other great powers were becoming more panicky and were hurrying for a decision, allowing Metternich to step in as the voice of calm and reason and point the way out. At Troppau, rather than admitting to self-contradiction in regards to Austria's contrasting views of Spain and Naples, he argued that while the Congress of Vienna had supported the protection the independence of all European states, and their complete freedom in its domestic arrangements, this freedom was nevertheless limited by the possible adverse international repercussions of their domestic actions. Thus, if a revolutionary government endangered the social order of its neighbors, this was no longer a purely domestic action, but a disturbance of international peace: in such context, foreign intervention was warranted. The revolutionary state was no longer an effective part of the Holy Alliance, and, as a result, was a legitimate target for intervention. As Gentz put it, the revolutionary principle represented

> a common enemy to all governments in office, as a fire which … anyone believes himself authorized and called upon to extinguish in his neighbor's home to avoid being oneself hurt, without claiming in so doing to meddle in what is going on in that neighbor's home or family.[39]

Of course, Metternich laid the case that, while Spain was not a danger to its neighbors, Naples was in precisely such an instance. By then, he had been lobbying directly the monarchs of Austria and Russia, bypassing their representatives, frightening the two kings for years on end with stories of a European-wide revolutionary conspiracy perpetuated by a directing committee located in France and with secret society cells in every major country. It was all too easy to portray every isolated protest or uprising in Europe as indications of further advances by the nefarious committee. The murder of a government publicist in Prussia by a student and a soldier rebellion among the Tsar's elite guard regiment was sufficient to push the two alarmed kings to give credence to Metternich's dire warnings. By the end of the Congress at Troppau, both were relying more on Metternich than on their own advisors.[40] A chastened Prussian King told Metternich that he had his full confidence because he had warned him, and everything he had foreseen had occurred. The Tsar went further, apologizing to Metternich for having mistrusted him in the past, and expressing the view that Russia was engaged in combat with the realm of Satan, by which he meant revolution. As for Britain, although it could not officially sanction Austrian intervention in Naples, being reluctant to endorse authoritarian principles over democratic ones, it unofficially approved of the intervention—it even suggested that Austria should proceed unilaterally. Metternich was to comment on the stance of his British counterpart that he was like a music-lover in church: he would have loved to applaud the good music being played, but did not dare to. In the end, the Congress not only empowered Austria to take action to repress the revolution in Naples, but also it endorsed its interpretation of the doctrine of intervention in case of international crises caused by revolutions. But by so doing, the great powers were acknowledging that revolutionary movements were effectively a pan-European problem, and, as such, not just Austria's concern but their own as well.[41]

Metternich's diplomatic triumph was completed by the change of course Russia made over Greece, at the time a province of the Ottoman Empire. Since the days of Peter the Great, Russia had been interested in expanding at the Ottoman Empire's expense. But if the Ottoman Empire collapsed, a new scramble to divide its remains would have started anew among the great powers (France already was preparing for just such a development to make a renewed bid for Belgium). It was therefore Austria's interest to preserve the Ottoman Empire, the so-called "sick old man of Europe," so as to avoid the prospect of a general war. However, in 1821, a rebellion flared up in Greece, which led to a massive effort of repression at the hands of the Sultan, culminating in the hanging for treason of the Patriarch of Constantinople from the door of his own church. Obviously, this represented a golden opportunity for Russia to step in under the pretext of saving its co-religionaries. Yet, Metternich managed the impressive feat of making Russia abandon its project, by convincing it, in the words of Kissinger, that it did not want after all to carry out the intervention. Metternich again successfully lobbied the Tsar: since the Greek uprising against the Sultan was but another manifestation of revolutionary spirit against a legitimate government, any Russian support would have benefitted the Paris directing committee. If any intervention should have occurred, it should have been against the Greek rebels to preserve the authority of the Sultan as the rightful sovereign. As Metternich wryly commented, it did not even matter that hundreds of thousands of Greeks were likely to end up beheaded or impaled, or that Russia had essentially disintegrated Peter the Great's foreign policy toward the Ottoman Empire.[42]

Metternich's domination of the international agenda was not to last. He well realized that skillful diplomacy could achieve only so much, when not supported by adequate military and economic resources. As he put it: "I have the feeling that I am in the middle of a web that I am spinning ... A net of this kind is good to behold, woven with artistry, and strong enough to withstand a light attack, even if it cannot survive a mighty gust of wind." Metternich's web depended on his carefully cultivated ability to persuade or to manipulate certain foreign decision-makers to follow the line he set out. With different people in charge, Metternich could not have counted on such a receptive audience. He had already acquired a reputation for duplicity that did not make the task of gaining the confidence of new decision-makers any easier. Accordingly, the suicide of the British foreign minister Castlereagh and the death of Tsar Alexander left Metternich to contend with different personalities, who distrusted his advice. As will be seen in the next case study, Great Britain had interests of its own outside Europe, and rejected to support Metternich's policy of cracking down on revolutions, which were, manifestly, not a British concern. Meanwhile, the new Tsar Nicholas I sought to reaffirm Russia's policy as an independent great power. He had a healthy suspicion of Metternich, whom he viewed as "the cohort of Satan," and surreptitiously crossed himself every time they had to meet. As a result, Austria's influence declined—there were to be no more congresses in which Metternich could charm and outfox the other representatives. In 1825, with the situation in Spain worsening, France managed to finally intervene to restore the royalist regime. This was followed by the conclusion of a joint understanding between Britain and Russia concerning the independence of Greece from the Ottoman Empire. Austria could only protest. Russia then proceeded to go to war with the Ottoman Empire. Despite Metternich's efforts, successful revolutions occurred in Belgium and France. In the 1830s and 1840s, every important international question was settled by decisions taken by Britain, Russia, and France, with Austria being confined to the sidelines.

And yet, an argument could be made that Metternich was still very much in charge. Even though Austria could no longer punch above its weight, the other great powers were still playing by the rulebook that Metternich had drawn, having internalized it to the point where it shaped their conceptions of national interest. For instance, even though it emerged victorious from the war with the Ottoman Empire, Russia confined its territorial demands to a patch of territory along the mouths of the Danube— even the cautious Nicholas now took for granted that it was better to keep the sick old man of Europe on life support, instead of giving him the coup de grace. Similarly, revolution in France in 1830 produced a conservative regime that promptly rejected sponsoring the cause of revolution abroad. All great powers accepted that preserving the Austrian Empire was necessary in order to maintain European peace and stability. None (with the arguable exception of France over Belgium) contemplated overthrowing the territorial settlement decided at Vienna, and risk general war. Even France, when presented with the chance of going to war in 1840, balked at the prospect. All great powers regularly consulted the others before pursuing any major undertaking. It is true that Britain and France opposed the principle of foreign intervention, but neither did they take action to promote liberalism abroad, not even in the context of the revolutions of 1848 in Poland, Hungary, Germany, and Italy, which were forcefully put down. To this extent, until the Crimean War and the advent of Bismarck, it would be fair to refer to the concert of Europe as the Metternich system.[43]

Notes

1 Jervis, "Unipolarity," 200.
2 Posen and Ross, "Competing Visions"; Jervis, "Remaking of a Unipolar World."
3 Posen and Ross, "Competing Visions," 32–5; Friedberg, *Weary Titan*, 144–52; Kagan and Kristol, "Towards a Neo-Reaganite Foreign Policy," 22, 26; Wohlforth, "Stability of a Unipolar World."
4 Friedberg, *Weary Titan*, 200.

5 Jervis, *Perception and Misperception*, 62–76.
6 Crawford, "Preventing Enemies Coalitions," 156–62; Eric Posner, Kathryn Spier, and Adrian Vermeule, "Divide and Conquer," *Coase-Sandor Working Paper Series in Law and Economics*, at https://chicagounbound.uchicago.edu/cgi/viewcontent.cgi?article=1 170&context=law_and_economics, esp. 29–31.
7 Holbraad, *Concert of Europe*, 3–4.
8 Jervis, "From Balance to Concert," 60–6. For a useful definition of engagement see Resnick, "Defining Engagement," 559–60; Victor Cha, "Engaging North Korea"; Johnston, *Social States*.
9 Gentz, *Dépêches*, 354–5; Holbraad, *Concert of Europe*, 18–9.
10 Schroeder, "The 19th Century International System," 12–13. Also see for a study explicitly linking prospect theory and French grand strategy under Louis XIV, Lynn, *Wars of Louis XIV*, 43–4.
11 As Schroeder emphasizes, the five great powers held no comparable capabilities or status in the settlement at Vienna. Schroeder, "Nineteenth Century System"; Elrod, "The Concert of Europe," 166–7.
12 Ibid., 164–5.
13 Jervis, "From Balance to Concert," 61–2.
14 McKay and Scott, *Rise of the Great Powers*, 14–16; Lynn, *Wars of Louis XIV*, chaps. 2–3; Bluche, *Louis XIV*, 221–32, 297–302, 435; Meyer, "Louis XIV et les puissances maritimes"; Angus Maddison, "Statistics on World Population, GDP and Per Capita GDP, 1–2008 AD", at http:// www.ggdc.net/maddison/; Zeller, "French Diplomacy and Foreign Policy," 198–9, 206; Lossky, "International Relations in Europe," in ibid., 178–82.
15 Church, *Richelieu*, 284–8, 295–8; Elliott, *Richelieu and Olivares*, 99, 122–4, 125–8; Zeller, "La Politique des frontières," 104–5; Erlanger, *Richelieu*, 453–4; Iskander Rehman, "Raison d'État: Richelieu's Grand Strategy During the Thirty Years' War," *Texas National Security Review* 2, no. 3 (2019), available online June 2019; Kissinger, *Diplomacy*, 56–77.
16 Bertière, *Mazarin*; Goubert, *Mazarin*.
17 Lossky, "International Relations in Europe," 189. Status concerns figured prominently in Louis' interest in consolidating France's position. Zeller writes for instance that "the quest for glory, then, took place of a programme for Louis XIV." Zeller, "French Diplomacy and Foreign Policy," 209, 206–8; Lynn, *Wars of Louis XIV*, 30–2; Blanning, *Pursuit of Glory*.
18 Lynn, *Wars of Louis XIV*, 35–7, 75; Blanchard, *Vauban*, 197–201; Bluche, *Louis XIV*, 286–9. As Derek McKay and H. M. Scott observe, the grand strategy of Louis XIV was one of limited objectives, not a quest for European hegemony on a Napoleonic or Hitlerite scale. McKay and Scott, *Rise of Great Powers*, 16–17; Bluche, *Louis XIV*, 283–6; Lynn, *Wars of Louis XIV*, 41–2. Also, the continuous concern of French leaders from Richelieu to Louis XIV for changes of borders should not be understood as a demand for "the natural frontiers of France," an idea that did not exist at the time, not in the least because their interests extended beyond natural limits such as the Rhine and the Alps. Zeller, "French Diplomacy and Foreign Policy."
19 Lynn, *Wars of Louis XIV*, 105–8.
20 Sonnino, *Louis XIV and the Origins of the Dutch War*; Bluche, *Louis XIV*, 246–8.
21 Lossky, "International Relations in Europe," 183–4; Lynn, *Wars of Louis XIV*, 109–11.
22 Ibid., 113–59; Bluche, *Louis XIV*, 246–63.
23 Ibid., 286–93; Lynn, *Wars of Louis XIV*, 161–4; Rule, "Louis XIV, Roi Bureaucrate," 68–9.
24 Another consideration was Louis' persecution of Protestants. In 1685 he revoked the Édit of Nantes, which had allowed for decades the Huguenots freedom of public worship in France, forcing hundreds of thousands into exile, most of them into the service

of French enemies. As a result, he was now reviled both by Catholics for siding with the Turks, and by Protestants for his crackdown. Lynn, *Wars of Louis XIV*, 174–9.

25 Ibid., 165–6, 191–3; Bluche, *Louis XIV*, 419–24; Wolf, *Emergence of the Great Powers*, 20–5.

26 Ibid., 39–42; Lynn, *Wars of Louis XIV*, 192–263; Bluche, *Louis XIV*, 422–41.

27 Wolf, *Emergence of the Great Powers*, 59–63; Lynn, *Wars of Louis XIV*, 266–70; Bluche, *Louis XIV*, 514–26. Louis kept underestimating the effect of security dilemmas on other great powers. England and Holland initially went ahead with the choice of Philip as King of Spain, but Louis in his desire to prepare the best he could for what he believed to be an inevitable war, pressed his advantage further by seizing fortifications that belonged to Spain in the Spanish Netherlands, in the process ejecting the Dutch forces that had held them since Ryswick, and by refusing to evict James's son, the Old Pretender, who still had claims on the British crown. Rule, "Louis XIV, Roi Bureaucrate," 83–4.

28 Lynn, *Wars of Louis XIV*, 271–32, 326–7; Bluche, *Louis XIV*, 626, 526–43.

29 Ibid., 543–72; Lynn, *Wars of Louis XIV*, 359, 328–58.

30 Philip V was fifth in line for the French crown when the war started. The only reason that there was a realistic chance of him becoming French king was the impossible to foresee successive losses of Louis' son and two grandsons in 1711. Nonetheless, the French throne went then to a great-grandson, the future Louis XV. True, Louis refused to admit that Philip did not retain any rights on the French crown, but, as the Spanish will required Philip to reside in Madrid, and, since, at the time, four other persons existed between him and the succession, this was a moot matter. Louis only rejected any renunciation because it interfered with what he held as divinely given rights of succession. Ibid., 361.

31 Lynn considers Louis XIV's grand strategy as bumbling into needless wars that sapped away French strength, particularly in the War of the League of Augsburg. However, the reduction of the French advantage compared to other great powers, in particular Britain, and to a lesser extent Russia and Austria, cannot be traced simply to the ruinous effects of Louis' continual wars. Endless warmongering was very much the fashion of the time, as seen in the cases of Peter and Frederick the Great. Instead, the rise of competitors was due to the structural shift of capabilities, especially economic and technological, or to gains made by other great powers in areas far away from France in Central Europe and the Baltics. France could have done little to prevent such developments. Lynn, "A Quest for Glory," 178–204, 200–1, 203–4; Bluche, *Louis XIV*, 627–9; Rule, "Louis XIV, Roi-Bureaucrate," 89–90.

32 Gentz, *Dépêches* I, 360–1; Sked, *Decline and Fall of the Habsburg Empire*, 1–2, 13–15; Kennedy, *Rise and Fall of Great Powers*, 149, 154, 162–5, 174; Maddison, "Statistics on World Population, GDP and Per Capita GDP."

33 Bridge, *Habsburg Monarchy*, 28–9; Kissinger, *World Restored*, 7; Kissinger, *Diplomacy*, 85–8.

34 Sked, *Decline and Fall*, 10–1; Kissinger, *World Restored*.

35 As Kissinger sums up: "Metternich's policy had ... taken the only form by which a state, aware of its weakness, can preserve the status quo without exhausting its resources: the creation of a moral consensus." Kissinger, *World Restored*, 269, 321.

36 Up to the dying stages of the War of 1814, Metternich had sought to strike a peace deal with Napoleon who was connected by marriage to the Habsburg imperial family, and could thus maintain in being a Franco-Austrian alliance. See ibid.

37 This lingering suspicion of French intentions and government stability was not unjustified. In the middle of the Congress, Napoleon escaped Elba, was promptly restored to power in Paris, and the allies had to again wage war on him, ending in his final defeat at Waterloo. This led to a second peace treaty with France, which placed part of the country under foreign occupation until 1818.

38 Kissinger, *World Restored*, 155–72; Bertier, *Metternich*, 236–46; Palmer, *Metternich*, 131–43; also see for a different interpretation that focuses more on the role of the British foreign minister Castlereagh, Schroeder, *Transformation of European Politics*, 523–38.

39 Kissinger, *World Restored*, 248–61; Schroeder, *Metternich's Diplomacy*, 60–94; Bertier, *Metternich*, 334–6, 338–43; Palmer, *Metternich*, 193–8; Gentz, *Dépêches* II, 195–6.

40 Accordingly, it could be said that Metternich had effectively supplanted the prime minister of each state. Kissinger, *World Restored*, 238–44, 259–60, 281, 302–4.

41 As Kissinger writes, "confronted by the prospect of nationalism and liberalism, he succeeded in making it a European, rather than an Austrian contest." Kissinger, *World Restored*, 266–7, 260–6; Schroeder, *Metternich's Diplomacy*, 43–4, 86–7. True enough, Britain and France refused to sign the Troppau Congress protocol, turning the Holy Alliance into a combination of Russia, Prussia, and Austria, but they tacitly accepted its consequences.

42 Kissinger, *World Restored*, 286–93, 298–309; Schroeder, *Metternich's Diplomacy*, 165–93; Bertier, *Metternich*, 354–5, 360–2.

43 Sked, *Metternich and Austria*, 65–73.

Bibliography

Bertier de Sauvigny, Guillaume. *Metternich*. Paris: Fayard, 1986.

Bertière, Simone. *Mazarin: Le Maître du Jeu*. Paris: Libraire Générale Française, 2016.

Blanchard, Anne. *Vauban*. Paris: Fayard, 2007.

Blanning, Tim. *The Pursuit of Glory: Europe 1648–1815*. New York: Penguin Books, 2008.

Bluche, François. *Louis XIV*. Oxford: Basil Blackwell, 1984.

Bridge, FR. *The Habsburg Monarchy Among Great Powers, 1815–1918*. Oxford: Berg, 1990.

Cha, Victor. "Engaging North Korea Credibly." *Survival* 42, no. 2 (2000): 136–55.

Church, William. *Richelieu and Reason of State*. Princeton: Princeton University Press, 2015.

Crawford, Timothy. "Preventing Enemies Coalitions: How Wedge Strategies Shape Power Politics." *International Security* 35 (Spring 2011): 155–89.

Elliott, JH. *Richelieu and Olivares*. New York: Cambridge University Press, 1991.

Elrod, Richard. "The Concert of Europe: A Fresh Look at an International System." *World Politics* 28 (January 1976): 159–74.

Erlanger, Philippe. *Richelieu*. Paris: Perrin, 2016.

Friedberg, Aaron. *Weary Titan: Britain and the Experience of Relative Decline, 1895–1905*. Princeton: Princeton University Press, 1988.

Gentz, Friederich. *Dépêches Inédites aux Hospodars de Valachie, I*. Paris: Plon, 1876.

Goubert, Pierre. *Mazarin*. Paris: Fayard, 1990.

Holbraad, Carsten. *The Concert of Europe: A Study in German and British International Theory, 1815–1914*. Harlow: Longmans, 1970.

Jervis, Robert. "From Balance to Concert: A Study of International Security Cooperation." *World Politics* 38 (October 1985): 58–79.

Jervis, Robert. *Perception and Misperception in World Politics*. Princeton: Princeton University Press, 1975.

Jervis, Robert. "The Remaking of a Unipolar World." *Washington Quarterly* 29 (Summer 2006): 7–19.

Jervis, Robert. "Unipolarity: A Structural Perspective." *World Politics* 61 (January 2009): 188–213.

Johnston, Alastair Iain. *Social States: China in International Institutions, 1980–2000.* Princeton: Princeton University Press, 2008.

Kagan, Robert and William Kristol. "Towards a Neo-Reaganite Foreign Policy." *Foreign Affairs* 75 (July/August 1996): 18–32.

Kennedy, Paul. *The Rise and Fall of Great Powers: Economic Change and Military Contest from 1500 to 2000.* New York: Random House, 1987.

Kissinger, Henry. *A World Restored: Metternich, Castlereagh, and the Problems of Peace.* Boston: Houghton Mifflin, 1957.

Kissinger, Henry. *Diplomacy.* New York: Simon & Schuster, 1994.

Lossky, Andrew. "International Relations in Europe." In *The New Cambridge Modern History,* edited by FL Carsten, vol. V. Cambridge: Cambridge University Press, 1961.

Lynn, John. "A Quest for Glory: The Formation of Strategy Under Louis XIV, 1661–1715." In *The Making of Grand Strategy: Rulers, States, and War,* edited by Williamson Murray, MacGregor Knox, and Alvin Bernstein. Cambridge: Cambridge University Press, 1994.

Lynn, John. *The Wars of Louis XIV.* New York: Routledge, 1999.

McKay, Derek and HM Scott. *The Rise of the Great Powers, 1648–1815.* New York: Longman, 1983.

Meyer, Jean. "Louis XIV et les puissances maritimes." *XVIIe Siècle* 31 (1979): 115–72.

Posen, Barry and Andrew Ross. "Competing Visions for US Grand Strategy." *International Security* 21 (Winter 1996): 5–53.

Palmer, Alan. *Metternich: A Biography.* New York: Harper & Row, 1972.

Resnick, Evan. "Defining Engagement." *Journal of International Affairs* 54 (Spring 2001): 551–66.

Rule, John C. "Louis XIV, Roi Bureaucrate". In *Louis XIV and the Craft of Kingship,* edited by John C Rule. Columbus: Ohio University Press, 1969.

Schroeder, Paul. *Metternich's Diplomacy at Its Zenith, 1820–1823.* New York: Greenwood Press, 1969.

Schroeder, Paul. "The 19th Century International System: Changes in Structure." *World Politics* 39 (October 1986): 1–26.

Schroeder, Paul. "The Nineteenth Century System: Balance of Power or Political Equilibrium." *Review of International Studies* 15 (1989): 135–53.

Schroeder, Paul. *The Transformation of European Politics, 1763–1848.* New York: Oxford University Press, 1994.

Sked, Alan. *Metternich and Austria: An Evaluation.* New York: Palgrave Macmillan, 2008.

Sked, Alan. *The Decline and Fall of the Habsburg Empire, 1815–1918.* New York: Longman, 1989.

Sonnino, Paul. *Louis XIV and the Origins of the Dutch War.* New York: Cambridge University Press, 1988.

Wohlforth, William. "The Stability of a Unipolar World." *International Security* 24 (Summer 1999): 5–41.

Wolf, John B. *The Emergence of the Great Powers, 1685–1815.* New York: Harper, 1951.

Zeller, Gaston. "French Diplomacy and Foreign Policy in Their European Setting." In *The New Cambridge Modern History,* edited by FL Carsten, vol. V. Cambridge: Cambridge University Press, 1961.

Zeller, Gaston. "La Politique des frontières au temps de la prépondérance espagnole." *Revue Historique* 193, no. 2 (1942): 97–110.

5   The grand strategies of
status quo powers II

This chapter is concerned with the other two grand strategies employed by status quo powers: semi-detachment and containment. The chapter begins by presenting the main provisions of each grand strategy and then provides historical exemplifications.

Semi-detachment

Semi-detachment refers to a grand strategy that combines the pursuit of primacy in the system's periphery with limited involvement in the core. In this way, the status quo is upheld in both the core, where the great power does not have any ambition, and in the periphery, where it consolidates an already existing predominance. This grand strategy is best seen in the cases of Great Britain, throughout the nineteenth century, and, arguably, of the US since its formation and up to 1898. As Daniel Baugh writes, "the English grand strategy ... was essentially defensive in Europe (and European waters) and aggressive overseas."[1]

Semi-detachment is a grand strategy of two halves, each dependent on the other. On the one hand, maintaining primacy in the periphery is concerned, just as in the above-mentioned grand strategy of primacy, with making unassailable the power and status of the great power, which could be either a dominant state or an established power. Periphery denotes a perimeter that is both distant and separate from the core system where the club members are located. For Great Britain, it referred broadly to the world outside Europe; for the US, it meant the confines of the North American continent. Therefore, restricting primacy to the periphery makes it a very different endeavor from the full version of the grand strategy for several reasons. First, semi-detachment is cheaper, because by limiting involvement in the core, the semi-detached power saves sufficient resources to ensure an upper hand in the periphery. Second, it is less likely than primacy to generate opposition from the other great powers. Some members of the club may not have any interests in expanding in the periphery themselves, or no way to gain access to it, for instance, by being landlocked. Third, if a great power expands in the periphery, it does not present the others with security dilemmas, because expansion does not occur in their own neighborhood, or near their sphere of influence. Fourth, the other great powers may not be able to detach themselves from the problems of the core. As such, they are not able to amass enough resources to both hold their own in the core and compete in the periphery. The semi-detached power, however, is a state whose geographical features shelter it to a great degree from the attack by the other great powers. Great Britain is an island, and the US is flanked, as French Ambassador Jules Jusserand's adagio went, by fish on both east and west, and by weak neighbors on both north and south. Consequently, these insular states can afford to spend less on defending their borders than a country located in the midst of other great powers, such as France, Austria, and Germany.[2]

Yet, on the other hand, semi-detachment does not mean that the great power puts all its eggs in one basket, i.e., in the periphery. True, the great power's most important interests lie there, but all the actors that can impact those interests the most are to be found in the core. In this sense, as Kenneth Bourne remarks, Britain was sheltered, but not isolated from the European continent. Therefore, well-calibrated aloofness is required. The great power does not want to be bound by ties of alliance and partnerships or "entanglements" to the core powers, because it may be corralled into fighting to support them in the core for the sake of their interests, not for its own. But it cannot afford not to respond to all the developments in the core. There are three types of situations that require its intervention. First, the expansion of another great power in its immediate neighborhood; second, a club member gaining a foothold close to strategic chokepoints that would enable this power to compete in the periphery; third, a club member developing power projection capabilities rivaling its own. Thus, a semi-detached grand strategy requires alternating involvement with passivity in the core, while, at the same time, consolidating one's supreme position in the periphery.[3] Consequently, semi-detachment is best suited for a great power insulated from the core, both in the sense of having formidable natural defenses, and of being located nearest the periphery, so that it enjoys privileged access. Semi-detachment is also better suited for liberal democracies for two reasons. First, it responds to the public demand of spending more on butter than on guns by maintaining a less expensive military apparatus in the core than the other great powers. Second, liberal democracies are suspicious of a large military, which might be used to put in place a dictatorship or to curb individual freedoms. This consideration increases the appeal of a grand strategy that keeps a lid on military expenses and army size.

Semi-detachment is often referred to as either a naval or blue water grand strategy, or as offshore balancing. Yet, both of these alternative labels are problematic, since they are prone to cause misconceptions. Semi-detachment is more than just a naval or blue water strategy.[4] A purely naval grand strategy characterized by exclusive reliance on the navy—as summed up by the frequently mentioned motto that "it is upon the navy under the Providence of God that the safety, honor, and welfare of this realm chiefly attend"—has never existed as such.[5] True, semi-detachment is best practiced by island states. It also stands to reason that naval power is a significant element in this grand strategy because it allows trade access to the periphery, protects sea lanes, and restricts other great powers' ability to affect the periphery. Furthermore, the great power does not have to deploy a considerable army in the core, because it has fewer interests to advance there, and the navy is enough to defend the state against external invasion. But this being said, naval power would not be effective without the exercise of political control over the periphery, because it requires a global network of supportive bases and harbors for the purposes of repair, refueling, resupply, and refuge. This in turn necessitates acquiring an imperial, colonial, patronage, or alliance network, as well as the development of a sizable garrisoning army in order to protect it and maintain order throughout it. Even Alfred T. Mahan, the most famous proponent of navalism, made it clear that a naval strategy supposed a complementary expansionist foreign policy, aimed at obtaining and maintaining "healthy colonies." Hence, sea power is but one of the means in a larger strategy of primacy in the periphery, alongside bases, trade, finances, technology, and ground forces. As Paul Kennedy writes: "it was not by maritime methods alone, but by a judicious blending of both sea power and land power, that Britain rose to become the leading world power." In effect, Britain's strategy did not rely on the navy alone, but also on a large ground force stationed in the periphery, the Indian Army. While Britain's forces in the British Isles comprised only 19 battalions, those in India and the colonies counted 78. The Indian Army was indispensable for Britain to have any sort of empire, having fought in China, Ethiopia, Egypt, Sudan, Afghanistan, and Eastern Africa. As Lord Salisbury, prime minister from 1895 to 1902, commented, India was "an Eastern barrack in the oriental seas from which we may draw any number of troops without paying for them." Therefore, the British empire did not float on the navy alone.[6]

Meanwhile, offshore balancing, which will be discussed in Chapter Seven, refers to a grand strategy recommended for the contemporary US, which argues that a great power should sit out most great power squabbles, and intervene solely at critical junctures where the balance of power is under danger of being overthrown.[7] Nevertheless, offshore balancing only captures half of the grand strategy under consideration. That is to say that it neglects the periphery. Indeed, the great power pursuing semi-detachment does not employ the resources freed by its limited involvement in the core just so it can advance economic and social agendas at home, but instead proceeds to expand abroad in a sub-system where it faces no rival. In effect, by the end of the nineteenth century, Britain had acquired an empire of 12 million square miles encompassing a fourth of the global population; while, in less than a century, due to its steady expansion westward, a process known as Manifest Destiny, the US gained a land expanse stretching across the North American continent from the Atlantic to the Pacific.[8] Moreover, in its dealings with the periphery, the great power does not rely on limited involvement, but precisely on the sort of tactics which offshore balancing severely disavows: frequent use of force, expansion of its sphere of influence, and rejection of an equilibrium of capabilities. Even while championing such equilibrium in the core, the great power insists on preserving uncontested dominance in the periphery. As Quincy Wright noted, "each statesman considers the balance of power good for others but not for himself," while Nicholas Spykman believed that "the truth of the matter is that states are only interested in a balance that is in their favor. Not an equilibrium, but a generous margin is their objective."[9]

Semi-detachment has several downsides. The most common is a penchant for acquiring more commitments than the state has resources to defend. Part of the cause for imperial overstretch is the lack of a viable opposition in the periphery from another core power, which creates the opportunity to expand without fearing major repercussions; the other part is the so-called "turbulent frontier" dilemma. Gaining a strategic position in the periphery creates the need to defend it through gaining additional territory, which creates a new vulnerable position, which requires further expansion to defend properly, and so on.[10] The result is that the state ends up with far more territory and commitments than it requires. Also, as Lord Salisbury put it, semi-detachment is a "dangerous game to play," because it requires a skillful appreciation of precisely when to intervene and when to abstain from intervention in the core.[11] If the great power remains passive in the wrong circumstances, it may allow an opponent to become so formidable that it can be checked only at great cost. Moreover, the other great powers may grow so accustomed to the great power's laissez-faire attitude that they discount it in their calculations. Yet, if the great power acts too precipitately in circumstances that do not warrant intervention, it may lose credibility by having its threats exposed as a bluff. Even worse, it may end up igniting an unnecessary and ruinous war. As Thomas Schelling writes, "if one side yields on a series of issues, when the matters at stake are not critical, it may be difficult to communicate to the other just when a vital issue has been reached." Thus, a party that usually abstains from intervention may not be taken seriously once it does decide to stand its ground.[12] Finally, semi-detachment functions best under conditions where there is a disconnection in the international system between core and periphery. This was the case since the end of the fifteenth century to the end of the nineteenth, when one dealt with two loosely interconnected sub-systems: the Old World (Europe) and the New World (the Americas), or more exactly the world system outside Europe. But if the two sub-systems are brought closer by improvements in transportation and communications, it will be problematic to pursue different approaches in each. This makes implementing semi-detachment increasingly challenging in an era of globalization and of military technologies such as the intercontinental missile and the strategic bomber.

Containment

Containment is a grand strategy practiced by either dominant states or established powers whose goal is to prevent a specific weaker competitor from changing existing power and status relations in its favor. Great powers often acquiesce to rising powers securing gains, on condition that they do not affect significantly an overall distribution of power and status that benefits them. But if the aggrandizement of another great power leaves them in the worse spot, they are likely to resist it. Hence, containment is designed to fend off the challenge of a rising power to one's privileged position. As the label suggests, containment is not about inflicting losses on the opponent, or about reverting its gains, but about denying it further gains. The challenger is contained: as it cannot advance, it is stuck in a perpetual position of inferiority vis-à-vis the power being challenged. The goal of this grand strategy is eventually for the challenger to realize the futility of the challenge and to acquiesce to a diplomatic settlement that upholds the status quo. Containment is most in evidence in the grand strategy of the US in the Cold War.

Containment is often used interchangeably with the balance of power. However, balance of power remains a highly ambiguous and debated concept. One survey chartered eight different meanings, some of which were contradictory. Nevertheless, the most often used meaning refers to the preservation of power equilibrium, in the sense of avoiding the concentration of capabilities in the hands of a single power. Thus, balance of power is essentially about preventing hegemony, and therefore can be summed up as saying that "the emergence of any single state with the potential to dominate the international system will generate a blocking coalition of other powers." This, however, is not containment. The reason is that, in the balance of power, the target of balancing is the stronger party, therefore the one enjoying the advantage. The weaker club members have to join forces in order to prevent it from becoming even more powerful than it already is. Things are different in containment: the status quo power is actually the one with the advantage. Although its lead on the challenger is eroding, it is still enjoying the upper hand. This makes containment a grand strategy practiced from a position of preponderance. This position may not be uncontestable as in primacy, but it enables the power to block the path of the ascension of its challenger, without having to call into being a grand coalition, as in the balance of power. As the State Department's Policy Planning Staff concluded during the Cold War, for the US "to seek less than preponderant power would be to opt for defeat. Preponderant power must be the object of US policy."[13]

Containment relies on military, economic, diplomatic, and psychological tactics. Military containment is about deterrence, the tactic of threatening the opponent in order to prevent it from initiating an action harmful to your interests. Notably, containment does not seek to go to war against the challenger. War is actually a last resort and its occurrence marks the failure of deterrence: as Vegetius put it, *si vis pacem, para bellum* (if you want peace, prepare for war). If a challenger seeks to advance further, the power practicing containment will confront the challenger with such heavy costs that aggression does not pay off. This is the basic difference between containment and primacy, or between deterrence, which is founded on punishment of the opponent, and defense, which is based on denying it gains. Primacy seeks to make the state's position unassailable, so that any attack or advance by a challenger will inevitably fail. Meanwhile, containment cannot guarantee the attack or advance will fail, but it can make sure that the power inflicts such pain on the challenger that the attempted attack turns into a self-defeating Pyrrhic victory. This means that containment provides two sizable advantages. First, the power using containment does not have to maintain absolute military supremacy, or seize and maintain control over strategic chokepoints. All that is required is making threats and meaning them. Second, deterrence is reactive. It depends on the challenger whether it decides to initiate action or not. If it does not, as deterrence gambles, the threat will never have to be actually carried out. This makes containment a far cheaper grand strategy than primacy. As John Foster Dulles, President Eisenhower's Secretary of State put it:

> We keep locks on our doors, but we do not have an armed guard in every home. We rely principally on a community security system so well equipped to punish any who would break in and steal that, in fact, any would-be aggressors are generally deterred. That is the way of getting maximum protection at minimum cost.

Ideally, containment does not involve bloodshed at all.[14]

Economic containment seeks to keep ahead of the opponent in economic and technological terms. Since money and hi-tech are the sinews of modern war, as long as a power conserves a healthy advantage in these areas relative to its competitor, any challenge has little chance of succeeding, because the challenger is reluctant to take on an enemy that is more powerful. Therefore, if this advantage can be carried on indefinitely, the challenger is not able to find any opportune moment to make its move, being permanently confined to second-best. Staying ahead depends on the evolution of the labor force, the state of technological research and innovation, the availability of capital and markets, and the international conditions of trade and investment. It depends, as well, on efforts to prevent the challenger from developing, through economic sanctions that restrict its access to foreign capital, markets, and technology. An important point to emphasize is that containment only monitors one's position vis-à-vis the challenger, not as primacy does, relative any conceivable combination of forces. In theory, a great power may fall behind a third party, provided it stays ahead of the challenger, or strike deals with other rising powers (for instance, the US and China in the 1970s, or Great Britain in relations with Japan and the US in the early 1900s) to cut down its expenses in resisting the principal opponent.[15] Accordingly, diplomatic containment is connected to economic containment in the sense that a challenger may attempt to make up for its deficiencies by securing alliances that would allow it to level the playing field. Diplomatic containment is about ensuring this does not happen, by either creating a larger, stronger, and wealthier combination, or/and by taking steps to persuade resource-significant states not to join the challenger's camp. This is not necessarily about completely isolating the challenger— if the only allies it manages to acquire are far away, poor, weak, and needy, they would not add to its power, but subtract from it.

Lastly, psychological containment involves perception. It does not only matter that the opponent fails to advance any further, but also, crucially, that it is not seen as advancing. If the challenger is perceived and perceives itself as winning, it risks creating a snowball effect. In the first place, the challenger will be emboldened by indications of defeat or weakness on the part of the defender (even signals as benign as courtesy or engaging in negotiations), and is likely to issue additional and greedier demands. The reason is that for deterrence to work, the challenger has to believe that the state has the necessary skill and resolve to carry out its threats, in other words that its threats are effective. Defeat in one context calls this assumption into question, thus making it harder for deterrence to be effective in other contexts. Thus, a defeat anywhere represents a defeat everywhere. In the second place, defeat may determine states whose territory or resources are in contention to switch sides, in the hope of either being spared depredation or of currying favor with the victor. This is the basis of the so-called domino theory, or "the expectation that defeat or retreat on one issue or in one area of the world is likely to produce ... more demands and defections from allies." This is why the task of psychological containment is to convey the message that momentum is and will stay on the side of the status quo defender. This is done in two ways. The containment power can increase the cost of defeat for itself—in the sense of burning down bridges behind it so that no retreat is possible.One way is to incur solemn and rigid pledges, which are politically very harmful for decision-makers to break, to stand firm in the face of challenges The other way is to actually resort to demonstrations of force, by intervening even "in the least significant, the least compelling, and the least rewarding cases." In so doing, " reaction should be disproportionate to the immediate prevision or the particular interest at stake." Indeed, such demonstrations of force send the signal that defeat is unacceptable; and, considering that the intervening power is willing to show resolve even over trifles or secondary interests, that it is all the more likely to do the same over essential ones. Hence, "the less the occasion, the greater the response."[16]

Containment's major downside consists in a perpetual and inevitable tension between strategic and psychological considerations. For containment to work optimally, it does not have to stymie the challenger's progress all over the map. Provided that the challenger is unable to expand its influence over important areas due to their location and capabilities, gaining additional territory, allies, or clients would not bring it closer to its goal, which is to supplant the status quo power. This means that from a strategic perspective, containment should be highly selective, being applied only to those states and areas that are worthwhile from a material standpoint, while showing indifference to the fate of the rest. But this manner of proceeding in turn is questionable from a psychological perspective: if a defeat anywhere leads to a cascade of subsequent defeats elsewhere, then the stakes of the overall contest overweigh the ones involved at the matter at hand. Losing any individual game brings therefore one closer to losing the match; and, hence, even if the game itself may not be worth the candle, one still is constrained to contest it. To this extent, every single contest is treated as if it were a vital interest. The consequence is that containment walks a tightrope between squandering resources by unnecessary intervention over minor provocations in areas of secondary importance, and risking collapse through the accumulation of perceived setbacks.[17]

Another downside of containment is that it takes a considerable time to work, while incurring heavy opportunity costs. Containment prioritizes stopping the advance of the challenger, but, in order to do so, interests must be sacrificed in relation to other powers, so as to avoid their joining the challenger, or have their antagonism eat away at the power's resources. In turn, this requires reaching compromises with these powers, which leaves it at a constant disadvantage. Furthermore, the benefits of containment may not be visible for a long time, while its costs are both tangible and immediate. Even if the challenger does not advance, it may not show signs of "mellowing," which translates into still more costs. This makes containment a grand strategy of playing the long game, hence of considerable patience. As the US Joint Chiefs of Staff pointed out in the Cold War, containment essentially comes down to "patience and firmness." Therefore, it is imperative that the domestic public be kept on board, as it may become tired of constant losses of blood and treasury in exchange for paltry returns.

Finally, the success of containment also depends on accurately gauging the proper conditions to implement it. Containment is not suitable to a power experiencing decline, because in this case, it is futile to try to stop further advances by the rising power, which has already managed to overtake the declining state. The position of strength that is vital for containment to work has already been lost. Containment therefore should be put in place *before* decline has set in. Yet, conversely, containment should not be implemented too early. If a great power does not face any challenge in the present and the foreseeable future, taking preventive action risks being interpreted as a threat by other great powers. In this manner, containment may become a self-fulfilling prophecy, a forecast that brings about a new behavior that makes the forecast come to pass. A threat did not exist originally, but after unnecessary containment is put in place, it emerges in its aftermath.[18] Thus, containment is best understood as a response to the emergence of a rival that has not supplanted the power yet in terms of capabilities and status, but is on course to do so. Under such circumstances, even if the intentions of the would-be challenger are presently benign, it may be wise to err on the side of caution and take precautions against allowing it to rise any further. As Eyre Crowe, the British Foreign Service official who made the case for a threat emerging from Imperial Germany, put it:

> It is quite possible that Germany does not have, or ever will, consciously cherish schemes of so subversive a nature. ... But this is not a matter in which England can safely run any risks. ... It would not be unjust to say that ambitious designs against one neighbor's are not as a rule openly proclaimed, and that therefore the absence of such proclamation, or even the profession of unlimited and universal political benevolence are not in themselves conclusive evidence for or against the existence of unpublished intentions.[19]

Semi-detachment: Great Britain and the status quo

From the defeat of Napoleon to the end of World War I, Great Britain was the dominant state. The sources of Britain's strength are well-documented. Its lead in the industrial revolution, which allowed it by the 1860s to account for more than half the world's production of iron, and for 40–45% of the global industrial output; its dominance of world finance as well as of trade, of which it controlled about one-fifth; its supremacy in shipping, as the owner of one-third of the world's merchant marine; the command of the commons; and its ownership of the world's largest colonial empire, eventually stretching over a fourth of the world's surface.

Yet, the question remains: how did a country that accounted for only 2% of the world population manage to remain the number one for nearly a century, while fighting only a single war against another club member (Russia in the Crimean War), and, for all intents and purposes, without the help of any ally?[20]

The answer lies in the grand strategy of semi-detachment adopted by Great Britain. To be sure, in the eighteenth century, Britain already had significant commercial, colonial, and naval interests, but it is only a century later that these elements congealed into a grand strategy, due to the exit of France as a rival outside Europe.[21] Post-1815, Britain enjoyed "a virtual monopoly among European powers of overseas colonies, and the virtual monopoly of world-wide naval power." Thus, Britain exerted a clear preponderance in the periphery, as the owner of large portions of India, Canada, Australia, South Africa, and the Caribbean, as well as by controlling access in and out of Europe.[22] Yet, it would have been all too easy for Britain to get sucked back into the affairs of Europe, as Metternich sought. Britain, however, pulled back.

The statesman who asserted a clear separation of core and periphery was George Canning, foreign minister between 1822 and 1827. Britain had been both a founding member of the Concert of Europe and a staunch supporter of Austria against Russia and France. Nevertheless, through successive congresses, Britain had grown increasingly uneasy with being roped in to put down revolution or to act as counterweight to the other great powers in places such as Central Europe or Italy. Neither of these tasks was fundamentally a British interest. Britain had no ambitions of its own on the continent. Its naval mastery meant that it would have been close to impossible for any other great power to launch an invasion, meaning that Britain was more secure than at any point in its history. Its geographical separation from the continent also meant that it was far less likely than the other great powers to suffer revolutionary contagion. Britain thought of itself as "walking the plank" between conservative monarchies, such as Austria and Russia, and radical democratic regimes, on the model of revolutionary France. Consequently, it had no stake in encouraging or inhibiting the spread of either ideology. Based on these considerations, why exactly should Britain have been concerned about the continent? The break with the continental powers eventually occurred at the Congress of Verona of 1822, where the great powers offered support to France's project to intervene in Spain, and Britain walked out in protest. Canning's view had always been that either the great powers were in agreement beforehand, which made congresses useless, or that they were not, which made them dangerous, by airing out and exacerbating disputes. After Verona, he had the perfect excuse to refuse ever participating again in a similar exercise. As he jubilantly put it, "things are getting back at a wholesome state again. Every nation for itself, and God for us all." Essentially, the great powers on the continent could no longer take advantage of Britain, which had regained its freedom of action.[23]

It is important to stress that this did not mean that Britain had pulled back from Europe out of petulance, once it found out it could not have its way over Spain. This interpretation would have implied that Britain had completely stopped caring about the core, surrendering in disgust all decisions into the hands of the four continental powers. Nothing of the sort. Canning made it clear that Britain still had significant interests left in Europe, even interests for the sake of which it was prepared to fight. Specifically, these concerned the security of Portugal, with which Britain had a long-standing, even though low-key alliance going back to the Middle Ages, and the independence of the former Spanish colonies in Latin America. If France took advantage of its military action in Spain to either invade Portugal, or to claim Latin America on the grounds that it rightfully belonged to Spain, it would have had to contend with Britain. In so doing, Canning wanted to make clear that developments in the core that touched on British preeminence in the periphery were still seen as important.[24] Moreover, keeping in mind that Britain's greatest strength was its navy, Canning realized that the best location to conduct intervention was in areas close to the sea. Countering France in Spain, for instance, was an exercise that would have stretched British capabilities, by requiring the dispatch and maintenance of a large ground expedition corps. Threatening such an intervention could only have been a bluff, and, as Canning put it, "a menace not intended to be executed is an engine which Great Britain could never condescend to employ." By contrast, denying French control over Latin America by using just sea power was easy and cheap. Thus, the principle advanced by Canning was not to tip the scales on the continent one way or another when intervening, but, instead, to intervene with "a commanding force" in those areas where naval power could be best brought on to bear. In this way, Britain carefully picked up those fights in the core that could impact the most its position in the periphery; and that it knew it could win on its own, without having to rely on an ally.

It is in this sense that Canning remarked to Parliament that he had "called the New World into existence to redress the balance of the Old." This famous statement was not about creating an alliance with the breakaway Latin American in order to balance France or the Holy Alliance in Europe. Instead, it meant that by recognizing the independence of these new states, Britain was asserting its supremacy in the periphery, and that it was excluding the other core powers from it. This interpretation is supported by the so-called Polignac Memorandum (1823) that Canning concluded with France, which explicitly mentions that Britain was not prepared to go "into a joint deliberation upon the subject of Spanish America upon an equal footing with other Powers, whose minds were less formed upon that question and whose interests were less implicated in the decision of it." Effectively, any action in the core that could lead to rivalry in the periphery was construed as an act of enmity toward Britain. As the Austrian ambassador reported to Metternich, the British message was clear: "The influence of the [Continental] powers ceased with the bounds of Europe."[25] Hence, from Canning onwards, Britain's grand strategy to maintain its number one position was based on two pillars: uncontested primacy in the periphery, and a skillful alternation of intervention and non-intervention in the core. This latter pillar was eventually dubbed as splendid isolation, splendid in the sense that Britain was not excluded from alliances with the other great powers, but kept aloof out of its own volition. Therefore, this aloofness was not due to weakness compared to the other powers, but precisely reflected the opposite, that Britain was the more powerful state, with far wider interests. As Benjamin Disraeli, prime minister from 1874 to 1880, argued:

> The abstention of England from any unnecessary interference in the affairs of Europe is the consequence not of her decline of power, but of her increased strength. England is no longer a mere European power; she is the metropolis of a great maritime empire, extending to the boundaries of the farthest oceans.[26]

In regards to the first half of the strategy, primacy in the periphery, the linchpin of the British position through the nineteenth century was its hold over India, "the jewel in the crown," a territory much larger than today's India, which eventually grew to comprise as well present-day Pakistan, Bangladesh, and, eventually, Myanmar. This larger India accounted for 85% of the entire surface of the British Empire, for 72% of its total population, for between a third and a fifth of its estimated GDP, for more than a third of British total military forces, and for the bulk of the ground forces in Asia. As India's Viceroy Lord Curzon summed up in 1901: "As long as we rule India, we are the greatest power in the world. If we lose it, we shall drop to the level of a third-rate power."[27] Therefore, the foremost condition for the preservation of British primacy in the periphery was to ensure the security of India, as well as of communications between Britain and India. For dealing with these issues, Britain relied on the same set of tactics as primacy: control over strategic "chokepoints", marked superiority in power projection capabilities over the next-in-line competitors, and finally, expansion in areas of interest for their resources and location on the map.

Accordingly, Britain started up by annexing "a string of islands and settlements" containing "some of the choicest strategic ports along the sea-lanes of the world," which also served the function of allowing it "a convenient base in every ocean of the world." The list of acquisitions comprised Singapore, on the passage from the Indian to the Pacific Oceans; Hong Kong, in the China Sea; the Falklands, near Cape Horn, hence on the passage from the Atlantic to the Pacific; Aden, on the Red Sea; Cyprus, Malta, and the Ionian Islands, in the Mediterranean; Lagos and Zanzibar, on each side of Africa; and Alexandria, to control the Suez Canal.[28] Second, as early as 1817, Lord Castlereagh, Canning's predecessor as foreign minister, had advocated that Britain should "keep up a navy equal to the navies of any two Powers that can be brought against us." Throughout the nineteenth century, Britain's navy often exceeded this margin, being larger than the next three or four great powers' combined fleets. Up to the 1890s, Germany's fleet was negligible, and so was Austria's. Russia's navy was severely hampered by the lack of a harbor that would be ice-free in winter, and by the fact that it could only exit the Black Sea through the Straits. This only left France as a serious competitor at sea. But Britain had enough warships to defeat any conceivable combination of France plus anyone else. Even when other great powers began large navy-building programs at the end of the nineteenth century, Britain originally still kept a considerable advantage over them, by defining safety as the Two-Power Standard by the

Naval Defense Act of 1889. According to the Act, the British navy would have, at any time, more battleships than the combined forces of the next two naval powers. It was telling that when proposing the measure, the First Lord of the Admiralty argued that this was nothing new, but only making formal "the old standard, which preceding governments had set before themselves."[29] Third, in order to adequately protect the harbors and coastlines it had seized, as well as India itself, Britain created further defenses, by enlarging its sphere of interest. This became a major concern from the 1830s on, as a threat to India emerged not from the sea, but from the land, in the form of Russia's expansion in Central Asia, which brought it close to Persia and Afghanistan.[30] To render pointless any Russian move on India, Britain moved to acquire additional real estate, a decision that eventually was applied not just to India, but all around. The consequence was that Britain ended up in control of a huge empire, consisting of "Egypt, Sudan, Somaliland, Kenya, Uganda, Rhodesia, Nyasaland, Bechuanaland, the Transvaal, the Orange Free State, most of present-day Ghana and Nigeria, Papua, North Borneo, Upper Burma, [and] several Malayan states." All in all, between 1837 and 1901, during the reign of Queen Victoria (1837–1901), Britain's empire gained no fewer than 44 colonies and 93 protectorates worldwide.

Such expansion was by all means excessive. As Lord Palmerston, multiple times British foreign minister between 1830 and 1850, put it, it was as if a man owned a house in London and one in York, and somehow felt he had to own all the inns on the way. Or as, nearly 60 years later, Prime-Minister Arthur Balfour noted: "Every time I come to a discussion, I find there is a new sphere we have got to guard, which is supposed to protect the gateways of India. Those gateways are getting away further and further from India." Many of the new acquisitions constituted, as Paul Kennedy writes, strategic liabilities rather than assets, a classic case of imperial overstretch that British decision-makers appeared unable to stop.[31] And this growing empire did generate considerable costs, as it could not be defended by the navy alone, but necessitated the use of ground forces. From 1815 to 1901, Britain undertook a staggering 65 wars, at the cost of more than 110,000 dead, the equivalent of twice the US casualties in Vietnam. As Byron Farwell writes: "There was not a single year in Queen Victoria's long reign in which somewhere in the world her soldiers were not fighting for her or for her empire … in Asia, Africa, Arabia, and elsewhere, British troops were engaged in almost constant combat," whether to conquer new territory, protect it from belligerent neighbors, or repress insurgencies. However, it should be added, while annexation may have been a suboptimal strategy from the point of view of the discrepancy of means and challenges, it was not a ruinous one. Britain emerged the victor in nearly all of its colonial wars, and, although it eventually lost its lead in industrial production to the US and Germany, it maintained itself as the number one in terms of trade, finance, shipping, and navy until World War I. If constant vigilance and a readiness to go to war anywhere in the world were the price for enjoying uncontested preeminence in the periphery, then it was deemed so much worth it, that Britain "paid it without qualms or regrets or very much thought."[32]

The second half of the grand strategy, or judicious involvement in the core, continued to follow Canning's principles. First, Britain intervened over issues touching upon British interests, either involving its own security or its preeminence in the periphery, but abstained from intervention in any other circumstances. Second, Britain intervened on the continent at points where Britain's superior naval power was likely to be the most effective, but kept out of those issues that could only be settled by relying on ground forces and alliances.

These principles were most in evidence in the Palmerston's tenure of office. Palmerston, probably together with Metternich and Bismarck, was the foremost diplomat of the age. Britain, Palmerston reasoned, was a great power that had no eternal allies or perpetual enemies. This statement is often taken to mean that Britain could change sides according to convenience. In fact, it was an injunction for standing aloof from other powers and their quarrels, by placing British interests above all else. As Palmerston actually put it:

> I hold with respect to alliances that England is a Power sufficiently strong, sufficiently powerful to steer her own course, and not to tie herself as an unnecessary appendage to the policy of any other Government. ... We have no eternal allies, and we have no perpetual enemies. Our interests are eternal, and those interests it is our duty to follow. ... If I might be allowed to express in one sentence the principle which I think ought to guide an English Minister, I would adopt the expression of Canning, and say that with every British Minister the interests of England ought to be the shibboleth of his policy.[33]

As a rule, Britain sat out European conflicts that did not directly affect either its security or its empire: it did not intervene for or against any of the revolutions that occurred in the 1830s and 1840s, and did not prevent the national reunifications of Italy and Germany, much as they otherwise upset the status quo of Europe.

This was far from being isolationism, because Palmerston sprang into frantic diplomatic, and, if need be, military action every time that British security or British dominance in the periphery, in other words Britain's own interests, as opposed to those of other core powers, were in jeopardy. For example, Britain constantly monitored the situation of Belgium. Situated closest to its shores, Belgium had, unlike France, the deep harbors to host the warships for an invasion. Since the days of Louis XIV, Britain had militated for the territory to be kept out of French hands. In 1815, Belgium had been conferred to the Kingdom of Netherlands. But in 1830, revolution started, and, in the subsequent fighting, the Protestant Dutch were forced out of Catholic Belgium. Naturally, Britain was concerned at the prospect of French intervention that might have morphed into annexation. Palmerston found a brilliant compromise solution, by creating Belgium, as an independent state under a minor German prince, and then by having all the club members formally give up any claim on Belgian territory and guarantee the new entity's neutrality. Therefore, Belgium became a buffer between France and the German powers, which by denying both these sides an advantage, worked to the great benefit of Britain. This was not, however, an easy result to deliver, and in a diplomatic tour de force, Palmerston had to alternate aligning, first with France, so as to deter the Holy Alliance from intervening; then with the Holy Alliance, so as to deter France from taking advantage; and, then back with France, so as to coerce the Netherlands to accept the great powers' decisions. When France sought to capitalize on the fact that its forces were deployed in Belgium by asking once more for a rectification of the border to its advantage, Palmerston threatened war. His threat was effective: France stayed confined to its 1815 borders.[34]

However, the single most important issue in Britain's on-again, off-again relationship with the core was the Eastern Question, the technical name given to the issue of the potential dissolution of the Ottoman Empire. The Ottoman Empire stood abreast perhaps the most significant "chokepoint" in the world: the twin Straits of the Bosphorus and the Dardanelles, which separate Europe from Asia, and also allow passage from the Black Sea to the Mediterranean. This was and still remains today a particularly important strategic position because of the geographical situation of Russia. Basically, if Russia ever gains control over the Straits, its Black Sea fleet, which, unlike most of Russia's navy that is ice-bound in winter, can be used year-round, would be able to deploy worldwide; moreover, Russian troops would have an easier access path into Asia, and, hence, back in the nineteenth century, toward British India. Thus, in order to safeguard its primacy in the periphery, Britain had to prevent the Ottoman Empire from collapsing, and giving way to Russian dominance.

When Palmerston assumed the foreign ministry, the situation in the Ottoman Empire was in free fall. In 1832, a rebellion had broken up under the power-ful governor of Egypt, Mehmed Ali. This created a major crisis, which unfor-tunately Britain sat out—the greatest mistake Britain had ever made according to Palmerston. Mehmed Ali was the protégé of France, who was lobbying for the great powers to carve up the Ottoman Empire in return for being allowed to annex Belgium. This meant that the isolated Sultan had no other option than to accept help from Russia. By that time, the Russian goal was no longer to conquer the Ottoman Empire, but rather to transform it into a protectorate or a client-state. In 1833, Russia determined the Sultan to sign the treaty of Unkiar-Skellesi, by which the two states concluded an "alliance," which allowed Russian forces to be deployed near the Straits to deter further advances by Mehmed Ali. In return, and considering Russia did not need any military help, the Ottoman Empire was pledging to close down the Straits if Russia ever came under attack. To go from this provision to allowing Russian warships unimpeded passage through the Straits there was but a step. Palmerston's paramount objective thus became to reverse Unkiar-Skellesi, while at the same time preventing France and its Egyptian client from taking advantage of the British-Russian dispute to make gains. Palmerston's first move was to start undermining the Russian position by reversing the British stance. From then on, Britain was set to intervene to help the Ottoman Empire if in trouble: the British ambassador in Istanbul was empowered to summon the British Mediterranean squadron to the Straits if there was any renewed threat.[35]

In 1840, the Sultan declared war on Mehmed Ali, who by then was control-ling a sizeable portion of the empire: besides Egypt, he also now lorded over Syria, Sudan, and the Arabian Peninsula. Predictably, the Ottoman army suffered ignominious defeat, the Ottoman navy promptly deserted to the rebel, and the Egyptian forces advanced toward Istanbul. This time around, Britain intervened in strength, skillfully bringing along Russia and Austria, who did not want to see the rise of a powerful French client in place of the Ottoman Empire. Mehmed Ali's advance was not only stopped; his rising power was cut down to size—he was to give back all his new conquests, return the deserter fleet, and accept both limits on his army and that Syria would return to Ottoman control after his death. Refusal to respond within ten days would have led to his loss of Syria, further delay would also have resulted in the loss of Egypt. In response, France threat-ened war: the humiliation of its client deprived it of its most valuable bargaining chip to effect territorial change in Western Europe. But the Palmerston-led coali-tion stood its ground. By November 1840, forced out of Syria by the British navy and a coalition army, Mehmed Ali had to bow to British terms. Palmerston played his hand so well that at the same time he managed to convince Russia to transform the Unkiar-Skellesi treaty into an international agreement. Russia was particularly cash-strapped at the time, and realized that it could not have renewed the treaty in the face of British opposition and increasing influence over the Sultan. Therefore, it had to accept Palmerston's suggested compromise, which was for all great pow-ers to recognize the Sultan's right to close the Straits to every navy in times of war in which the Ottoman Empire was involved. On the surface, this was simply keeping with the terms of Unkiar-Skellesi, as the subsequent Russian spin argued. But in reality, what the new agreement had done was to make sure that in case of war, the Russian navy could not count on its Black Sea force, a provision very much in the benefit of Britain. The Turkish right to close the Straits in time of war has been in force ever since.[36]

This did not mean that Palmerston was immune from severe errors of judg-ment in deciding when to intervene. As seen in Chapter Three, he attempted a bluff, threatening Prussia and Austria with intervention on behalf of Denmark over the Schleswig-Holstein crisis. Bismarck, as we have seen, was unimpressed by Palmerston's bluster to send in the British navy. To stop the offensive of the two German powers into Denmark, the navy was ineffective, and if the British Army showed up, Bismarck was reported to have quipped he would have had it arrested by the Prussian police. Britain had to resort to a humiliating climb down from its bluff, which led it to refrain from again meddling in German affairs for the next half-century. Britain's subsequent abstention of intervention in the Franco-German War produced the unification of Germany, creating in the process a potentially more dangerous competitor to Britain's number one position than either France or Russia.[37] Worse, Britain's semi-detached attitude had the effect that when it eventually threatened intervention, other great powers were left in doubt whether it meant to really go through with it. Britain hence encountered difficulties in making its threats credible, because its commitments to the core could never be iron-clad, as those of the other powers. This may have been a contributing factor if not a direct cause of wars, such as the Crimean War and even World War I, that might have been prevented by an unambiguous British pledge to intervene.[38]

Containment: the United States and the status quo

The US may have never had it, either before or since, as good as it did in the aftermath of World War II. Yet, paradoxically, at the very same time, it also confronted "undoubtedly the greatest task our diplomacy ever faced and probably greatest it will ever face." On the one hand, the US dominance in select power dimensions was much more lopsided than ever, the post-Cold War included. The US was the beating heart of the world's economy, accounting for one third of the world's global production, half of its manufacturing, and two thirds of its invested capital. Washington deployed the largest air force and navy in the world, and enjoyed a monopoly over nuclear weapons. The rest of the great power club was in remarkably poor shape. Defeated Germany and Japan were under military occupation; France and Great Britain had suffered tremendous devastation; most of the states in Europe had trouble providing food, heating, and electric power for their inhabitants. As Churchill described it, it was all at once "a rubble heap, a charnel house, a breeding ground for pestilence and hate." The Soviet Union, the other great winner of World War II, had suffered literal decimation, losing more than 20 million of its population in the war, and the economic damage it had suffered surpassed the combined national wealth of Britain and Germany. On the surface, no great power was in a fit state to compete, at least not openly, with the US. And why would it have done so, since the victors of World War II had fought the Axis side by side? It was perfectly conceivable that wartime cooperation would have continued into a form of neo-concert, as in US President Franklin Delano Roosevelt's vision of four great powers, acting as world policemen in maintaining order and security in their quarters of the world.[39]

However, at the same time, the Kremlin had secured sizable gains in territory at the expense of the Baltic Republics, Finland, Poland, Romania, and Germany. Now, its army outnumbered America's three to one, since the US had decommissioned its soldiers after victory, cutting down an army of 10 million to 1 million, while the Soviets kept a third of their troops under arms, thus gaining a marked superiority in terms of ground forces. As a result, Soviet influence extended deep into Germany, Central, and Eastern Europe, much of it still under the Red Army occupation. This influence was not just exerted over the foreign policy of the occupied countries, but over their domestic politics as well, through rigged elections, coups, bans of other political parties, jailing of political opponents, and staffing of the resulting client governments with Soviet advisors. Eastern Europe was considered a litmus test for the Soviets' likely designs elsewhere, with the risk of a Soviet takeover of Western Europe as well, through subversion, infiltration, or conquest. Beyond Europe, the Soviet Union could have targeted the Middle East and East Asia, a move already portended by Soviet coercion of Iran and notably Turkey (as in Tsarist times, with the aim to win control of the Straits).

Therefore, by adding the resources of these other areas to the ones already under its control, the Soviet Union could have not just matched, but also gained the upper hand against the US. As President Truman put it,

> if Western Europe were to fall to Soviet Russia, it would double the Soviet supply of coal and triple the Soviet supply of steel. ... And Soviet command of the free nations of Europe and Asia would confront us with military forces which we could never hope to equal.[40]

To sum up: the US was clearly on top, but unless it took action vis-à-vis further Soviet expansion, it might not have remained there for long.

Accordingly, the question was how best to deal with the Soviets. Both the traditional solutions of concert or spheres of influence were unfeasible. For a concert to work, the other great powers should have been able to pull their own weight, but it was becoming increasingly clear that neither Britain, nor France, nor China had the resources needed to play their assigned role of world police-men. Consequently, a large void of power opened in both Europe and Asia, which the Soviet Union sought to fill. Meanwhile, a spheres of influence agreement that would have portioned the world into exclusive American and Soviet spheres was unacceptable to the US public, because it was no longer tenable in an age of democracy, mass participation in politics, and self-determination to dispose of whole nations without allowing them a decisive say in their own fate, as had been the case at the Congress of Vienna. Even worse, to strike such a deal with the Communist USSR, the remaining champion of anti-democratic principles, would have been "to admit the irrelevance of the American domestic experience." Hence, the US never accepted as legitimate Soviet control of Eastern Europe, nor could it ever assent to it formally. While it is true that an agreement on percentages of influence had been concluded during World War II between Britain and the Soviet Union in Moscow (October 1944) , from the point of view of the US, this was considered to have been superseded by ulterior agreements, such as Yalta (February 1945.) In any case, the Moscow agreement had not authorized the Kremlin to completely eliminate the economic, diplomatic, and political influence of other great powers from their sphere, in other words, to have a completely "closed" sphere of influence, or to intervene in the composition of the domestic institutions of the states located within in its sphere. Thus, what good was it to conclude a sphere of influence agreement with the Kremlin, if the other side was not going to respect it? As a dejected Roosevelt admitted, the Soviet Union had broken every one of the promises it had made at Yalta.[41]

The solution to the Soviet problem was suggested by George Kennan, an American diplomat with wide experience in Soviet affairs.[42] Kennan argued his case first internally in the aptly called "Long Telegram" sent from Moscow in February 1946 and then to the general public in a series of public presentations, a course on grand strategy offered at the National War College, and, most famously, in an article published under pseudonym (Mr. X) in the *Foreign Affairs* journal in July 1947. In order to come with a response, the first thing Kennan did was to determine what goals the Soviet Union sought to achieve. For Kennan, the Soviet leaders were mainly moved by the belief of an irreconcilable antagonism between capitalism and communism. While communism may on occasion have compromised, or even concluded alliances with the capitalism system for tactical reasons, as in World War II, its long-term goal was its destruction. This antagonism and its outcome had been predicted as inevitable by Marx and by Lenin, but this "everyone was out to get you" worldview also had a particular resonance for the Soviet leadership, because it tapped into both the historical experience of insecurity of a Russia under constant threat from its neighbors, and in the personal insecurity of the Soviet leadership, under constant threat of being overthrown by an oppressed population. As a result, as Kennan assessed,

> we have here a political force committed fanatically to the belief that with the US there can be no modus vivendi, that it is desirable and necessary that the internal harmony of our society be destroyed, our traditional way of life be destroyed, the international authority of our state be broken, if Soviet power is to be secure.

Therefore, it did not matter that the US had no aggressive intentions toward the the Soviet Union, or that it was prepared to live alongside it, because of the Kremlin's enduring conviction that all capitalists, and the US by excellence as the capitalist-in-chief, were by definition its sworn enemies. As Kennan pithily wrote: "It is an undeniable privilege of every man to prove himself right in the thesis that the world is his enemy; for if he reiterates it frequently enough, and makes it the background of his conduct, he is bound eventually to be right."[43]

This did not mean that the Soviet Union was Nazi Germany come again. Hitler may have gone ahead and attacked any enemy regardless of the price he had to pay. The Soviet Union, however, was a rational opponent: it realized that it could not have conquered Western Europe, without triggering American nuclear retaliation. Moreover, the Kremlin was aware of just how much World War II had left it weakened. Accordingly, it would not have risked open war. But it could have achieved its goals in an oblique way, by taking advantage of the wide-scale poverty and dejection in select countries in the West to undermine democratic regimes and place in power favorable regimes. This could have been achieved through a wide array of tactics, such as "persuasion, intimidation, deceit, corruption, penetration, subversion, horse-trading, bluffing, psychological pressure, economic pressure, seduction, blackmail, theft, fraud, rape, battle, murder, and sudden death," and Kennan added that this should not have been mistaken for a complete list of Soviet tactics. Furthermore, the Kremlin would also have opened new wedges and exploited existing ones between Britain and the US, Western Europe and the US, and colonies and Western Europe, in order to turn them against each other, while the Soviets kept advancing surreptitiously. In a nutshell, the Soviet Union sought to take over without having to fire a single shot.[44]

The appropriate response was for Kennan a "long-term, patient but firm and vigilant containment." Containment was in Kennan's vision multi-tiered. In the first place, it depended on deterrence in as far as this was a question of keeping in check Soviet military aggressiveness, and of making the Soviets back off in case of a confrontation. As Kennan put it: "you have no idea how much it contributes to the general politeness and pleasantness of diplomacy to have a little quiet armed force in the background." Second, related to the first point, the US was actually stronger than the Kremlin, so the key to containment was to ensure that Washington maintained its edge in capabilities, and held its nerve in the face of Soviet intimidatory probes. Without ever being able to gain the upper hand, the Soviets would have to back off every time. To quote Kennan: "a preponderance of strength" among democracies is "the most peaceful of all the measures we can take short of war because the greater your strength, the less likelihood that you are ever going to have to use it." Third, given that military force was in most cases not in the cards, the key to successful containment was for Kennan foremost economic and diplomatic (for instance, Kennan did not think that NATO was really necessary, apart from reassuring the Europeans the US was not going to abandon them). Hence, if the US helped Europe to get back on its feet, the avenue for surreptitious communist advances would have been closed. Moreover, if the US demonstrated that it supported its Western partners and stood by them, the Soviet Union would have had trouble in driving wedges between them. Fourth, although the "X" article is notoriously ambiguous on this point (it mentions the US responding at a "series of constantly shifting geographical and political points,") Kennan did not advocate a global containment, but rather opposed the spread of Soviet influence in select areas, valuable for their industrial and technological resources. His list of such areas initially comprised the Atlantic community (Western Europe plus Scandinavia, Canada, and Latin America), the Middle East up to Iran, as well as Japan and the Philippines. A further version spoke of "only five centers of industrial and military power in the world which are important to us for the standpoint of national security": the US, Britain, Germany, the Soviet Union, and Japan. Therefore, as long as the US managed to keep Britain, Germany, and Japan out of the Kremlin's grasp, it would have remained ahead, no matter what other headways the Soviets would have made elsewhere. It was therefore useless to throw away resources on containing communism in places such as China: it was not a military industrial center, and, therefore, would not have affected the outcome of the systemic contest taking place. Nor should the US have attempted to rollback communism in Eastern Europe—apart from Eastern Germany, the area did not have resources that would have conferred a significant advantage on the Soviets. Thus, containment was not about depriving the Soviets of their gains, but making sure they could not obtain additional ones. Fifth, the expected result of successful containment was that, having failed repeatedly to break through, the Kremlin would gradually abandon its aggressive ideology and learn to live and let live with the capitalist world. To quote Kennan: "for no mystical Messianic movement—and particularly not that of the Kremlin—can face frustration indefinitely without eventually adjusting itself in one way or another to the logic of that state of affairs." Actually, Kennan went beyond this prediction, and, as it turned out, presciently raised the possibility that having failed in its challenge, and having lost its ostensible raison d'être, the communist system might collapse altogether: "the possibility remains ... that Soviet power, like the capitalist world of its conception, bears within it the seeds of its own decay, and that the sprouting of these seeds is well advanced."[45]

It is not that the US had not stood up to the Soviet Union before 1947. To some extent, efforts to prevent an extension of Soviet influence in Western Europe, Japan, and the Middle East had been on-going from 1945. But until Kennan's formulation, this resistance did not amount to a systemic response aimed at achieving any goal apart from making the Soviets climb down in each separate incident. As the future Secretary of State Dean Acheson wrote, it was like the Soviets were throwing rocks through the window, and the US was trying to plug the resulting hole with a newspaper, only to move on to the next hole. But in early 1947, Britain let the US know that it was no longer able to provide financial support to the Greek government under assault by Communist insurgents. Either the US replaced Britain in defending Greece, assuming responsibility for leading and marshalling system resistance against the Kremlin, or it risked a Communist advance in the Mediterranean. As the number three at the State Department William Clayton summed up the choice:

> The reins of world leadership are fast slipping from Britain's competent but now very weak hands. Those reins will be picked up either by the United States or by Russia. If by Russia, there will almost certainly be war in the next decade or sthe odds against us. If by the United States, war can almost certainly be prevented.

Consequently, in March 1947, the US put forward the Truman Doctrine, which advocated extending support to "free peoples who are resisting attempted subjugation by armed minorities or by outside pressures." The Truman Doctrine represented no less than a tidal wave in the US foreign policy because it committed the US to the defense of Europe, something that had been off-limits since 1776. Therefore, Truman had accomplished the task that had eluded Woodrow Wilson, whose projected League of Nations had failed in the aftermath of World War I, with terrible consequences for Atlantic security. By contrast, the US now clearly stated that it had vital interests in Europe, for the sake of which it was prepared to fight. This revolution was fully completed in 1949 with the constitution of NATO, which pledged American military assistance to ten European countries.[46]

The Truman Doctrine was followed by the announcement of the Marshall Plan in June 1947, in a commencement speech offered by Secretary of State George C. Marshall at Harvard University. The Marshall Plan was fundamentally a calculated gambit. First, its ostensible goal was to provide US economic aid for the recovery of the war-torn European economies. This was not a paltry effort, but a serious commitment, on par with the defense expenses of the US: between 1947 and 1950 Washington sent to Europe $12.4 billion worth of aid, equivalating with 12% of the total US budget. Second, the Marshall Plan would reinvigorate the

Western economies, undercutting the appeal of communism. As Kennan eloquently put it in one of his lectures, appeal to democratic values was not enough.

> We have freedom of elections, freedom of speech, freedom to live our life
> politically; but a great many people in this world will say that is not enough;
> we are tired; we are hungry; we are bewildered; to hell with freedom to elect
> somebody; to hell with freedom of speech; what we want is to be shown the
> way; we want to be guided.

Communism was providing an all-too-easy answer out of that conundrum. But
if conditions ameliorated, there was no longer any need for a messianic movement to show the way out. Third, the Marshall Plan would expose a stark contrast
between what was going on in the two halves of Europe under American and
Soviet control. Destitute themselves, the Soviets were unable to provide any alternative to the Marshall Plan for the sake of Eastern Europe; and they were, at any
rate, unwilling to do so—in fact, the Kremlin had proceeded to methodically pil-
lage the industrial infrastructure of the region, with whole production plants being
disassembled and sent over to Soviet territory. Finally, the Marshall Plan also
trapped the Soviets because, on the initiative of Kennan, the plan was proposed to
cover not only Western Europe, but also be extended to Eastern Europe and to the
Soviet Union itself. The Soviets could not accept this much-needed economic aid
without, at the same time, agreeing to economic and political conditions, as well
as close inspection by the US. This would have been completely unacceptable to
Stalin. But refusing would have exposed to the world that Soviet propaganda was
a lie. Between the two superpowers, the one that provided aid to the needy was the
US; meanwhile, the Soviet Union was the villain of the piece, preventing aid from
reaching those in need. How could then the Soviet Union claim it cared about the
best interests of West Europeans, based on how it treated the states already under
its control? As predicted, the Kremlin had no option but to walk out in sullen protest from the negotiations on the Marshall Plan in Paris in 1947.[47]

Despite its initial success, Kennan's version of containment soon ran into a
fundamental obstacle: how to keep containment restricted to certain regions, while
allowing others to fall to communism, and at the same time preserve domestic and
international support? By 1949, communism appeared, on the surface at least, to
be gaining steady ground: the Soviet Union tested a nuclear device in August,
nullifying the US monopoly over atomic weapons. Then, the Communists tri-
umphed in the Chinese Civil War in October, and concluded an alliance with the
Soviet Union in February of the next year, thus adding China and its huge human
resources to the Kremlin's camp. From the perspective of Kennan's logic, these
developments were not as worrisome as they appeared. The US was still well
ahead economically and technologically, and, in atomic weapons, it disposed of in
the early 1950s more than 50 warheads as compared to the Soviets who disposed
of fewer than five. On account of its poverty and backwardness, China represented more of a drain than an asset. But in order to be practical, grand strategy
must be sellable to a domestic public, which was asked to contribute billions

of dollars to the recovery of Europe and the revitalization of the US military. Furthermore, another consideration was the effect of US passivity on the Kremlin and the US allies. Failure to act to support non-communist forces in one context could be an indication of weak resolve in standing up to them in all other contexts. As Acheson put it in an antecedent of the domino theory, if one apple in a barrel is rotten (communist), it would infect the rest.[48]

It is fair to say that containment effectively split into a minimalist and a maximalist version. The minimalist version, first proposed by Kennan, and then translated into practice to various degrees by the administrations of Truman (pre-1950), Eisenhower, Nixon, Ford, and Carter proceeded from the assumption that deterrence and economic and technological assistance were sufficient to make containment work. As such, there was reduced need for direct military intervention except at key strategic points, or for large military spending. If confronted with a clear threat, the Soviets would always choose to back off. Moreover, the US depended a great deal on the support of allies, who should have taken responsibility as well for their defense. By contrast, the maximalist version, which was seen in practice during the tenure in office of the Truman (post-1950), Kennedy, Johnson, and Reagan administrations, was based on countering the Soviet Union at every point, strategic or not, where it sought to secure an advantage. Deterrence was not enough, and had to be supplemented by military intervention and by a sizable defense budget. Allies could not be trusted to do the job on their own. The Cold War was not just a battle over power and status, but also, and more importantly, a global struggle between freedom and totalitarianism.

Maximalism came to the front in 1949–1950, in particular in the National Security Council report, codified as NSC 68. Kennan by then had resigned from the Department of State on account of his differences with Acheson over commitments to East Asia, a project for re-unifying Germany, and the militarization of containment. Paul Nitze, Kennan's replacement, together with Acheson, had been the key influences in producing NSC 68. The document presented an alarmist view of global communism as a monolithic force under complete command by the Kremlin. Therefore, any victory for communism was effectively a victory for the Soviet Union. Moscow was increasing its military spending to the point where, should this trend have been allowed to continue, US military superiority, reliant on nuclear weapons and the air force, might have been jeopardized. Consequently, the threat from the Soviet Union was not just political and economic, as Kennan had argued, but also military, and in this respect the US and its allies were not doing enough: "the actual and potential capabilities of the United States, given a continuation of current and projected programs, will become less and less effective as a war deterrent." This meant that any localized communist attack against free institutions would have confronted the US with the choice of either global atomic war or retreat, which would have been interpreted by the Kremlin as a sign of irresoluteness and desperation, and, as such, as an invitation to press even harder and expand its offensive. Hence, as NSC 68 concluded, "the assault on free institutions is worldwide now, and in the context of the present polarization of power, a defeat of free institutions anywhere is a defeat everywhere." Thus, the US had to react to every single communist advance, not just to those in key strategic areas, as Kennan had argued: "*any* [my emphasis] substantial further extension of the area under the domination of the Kremlin would raise the possibility that no coalition adequate to confront the Kremlin with greater strength could be assembled." For instance, as NSC 68 argued, the loss of Czechoslovakia to communism in 1948

> was not in the measure of Czechoslovakia's importance to us. ... But when the integrity of the Czechoslovak institutions was destroyed, it was in the intangible scale of values that we registered a loss more damaging than the material loss we had already suffered.

Accordingly, the remedies sought by NSC 68 were foremost military: a substantial increase in US military spending; the extension of military and economic support to forces resisting communism, not just in key regions, but worldwide; as well as nuclear escalation by developing thermonuclear weapons, with a thousand-fold destructive capacity compared to fission-based weapons. This was pretty much turning Kennan's argument upside down—as NSC 68 put it, "without superior aggregate military strength ... a policy of 'containment' ... is no more than a policy of bluff." It is important to underline that while NSC 68 meant what it said, it was also purposefully exaggerating the threat from the Soviet military build-up, as well as raising the possibility of not just denying gains to communism, but also of optimistically rolling it back by reducing Soviet power "to limits which no longer constitute a threat to the peace, national independence, and stability of the world family of nations." This was done in order to scare the US government and public into supporting a large increase in military spending. As Acheson confided, the purpose of the document was "to so bludgeon the mass mind of 'top government' that not only could the President make a decision but that the decision could be carried out." Edward Barrett, the Assistant Secretary of State for Public Affairs put it more bluntly: this was a "psychological 'scare campaign.'" This was not the first time either that the US government had resorted to such tactics: as Senator Arthur Vandenberg put it of the Truman Doctrine, its goal had been to "scare the hell out of the American people." In effect, the adoption of NSC 68 resulted in the eventual tripling of the US defense budget, from $17 billion to $52 billion per year, or from 4% of the yearly budget to 13%, a development that would have been fought tooth and nail by the bureaucracy, Congress, and public opinion, had they not been terrified of the prospect of communist victory.[49] It is telling that once the document was formally endorsed, the US made no actual attempt to roll back the Soviet power and influence in Eastern Europe or East Asia.[50]

The clinching argument for NSC 68, however, was, beyond anything that the US government might have said, the actions of the communist bloc, specifically the invasion of South Korea by North Korean forces in June 1950. As Acheson concluded, "Korea saved us." The end of World War II had left the peninsula divided at the 38th parallel into a northern communist and a southern pro-West state. However, in accordance to Kennan's vision, the US had no strategic interest in South Korea, something that Acheson unwittingly communicated in a press briefing in January 1950, in which he outlined the US security perimeter in East Asia, only mentioning the Philippines, Japan, Okinawa, and the Aleutians. Hence, from a purely strategic perspective, the US should have been unconcerned if Korea had been reunified under communist auspices. But from a psychological perspective, this was a very different picture. The invasion of South Korea could not have gone ahead without a go-ahead from the Kremlin (which, as was later proven by archival evidence, was a correct assessment). This made the invasion a communist probe aimed to sound the Western will to resist. If the US let communism win in Korea, new probes would have been attempted against Western Europe's defenses, and doubts would have surfaced among European allies as to the reliability of the US as an ally. Accordingly, for President Truman,

> in my generation this was not the first time when the strong had attacked the weak. I recalled some earlier instances: Manchuria, Ethiopia, Austria. I remembered how each time that the democracies had failed to act it had encouraged the aggressors to keep going ahead. Communism was acting in Korea just as Hitler, Mussolini, and the Japanese had acted ten, fifteen, and twenty years earlier.

Fearing a third world war, Truman reasoned that "if we let Korea down the Soviet will keep right on going and swallow up one piece of Asia after another. ... Then the Near East would collapse and no telling what would happen in Europe." Acheson, meanwhile, thought that abandoning Korea would make the Americans "the greatest appeasers of all time." Faced with what appeared as unmistakable confirmation of Communist aggressiveness in Korea, the US leaders and public took the threat seriously, and, in so doing, subscribed to the logic of NSC 68: a defeat anywhere had become a defeat everywhere. As the *New York Times* stated, "we can lose half the world at this point, if we lose heart." Even Kennan supported intervention.[51]

Containment is considered a clear example of a successful grand strategy. Kennan's own view after the end of the Cold War was far more nuanced: in its maximalist version, containment had become far too militarized, downplayed the role of diplomacy, and involved the US in a series of interventions in places of little actual strategic significance for deciding the outcome of the Cold War. Effectively, by transforming the contest between the US and the Soviet Union into a global confrontation, containment ended up being excessively costly and risky. The US lost 55,000 soldiers in the Vietnam War alone; spent trillions in developing a lopsided conventional and nuclear arsenal (at its height in the mid-1960s owning an excess of 31,000 warheads); and spent inordinate amounts of resources in combating left-leaning governments in Southeast Asia, Central America, sub-Saharan Africa, and the Middle East. In the cases of Iran, Guatemala, Brazil, and Chile, on the grounds of resisting the advance of communism, the US played a key role in overthrowing legitimately elected governments. Furthermore, the US risked nuclear war with a potential death toll of 100 million Americans killed over protecting its reputation in places such as Quemoy and Matsu (Taiwan), Berlin, and Cuba. In this last instance, in which the US challenged the Soviet Union to remove nuclear missiles it had placed in communist-held Cuba in October 1962 despite previous assurances it would not resort to such a move, President Kennedy thought that the chances of war were between one-in-three and even. Yet, it should be said that maximalist containment for all its costs had also two positive results. First, the Kremlin was very rarely perceived to be making serious inroads and the consensus view all throughout the Cold War, including from the Soviet side, remained that the US was the stronger party. Second, America's global commitment eventually forced the Soviets to follow suit. Yet, the Soviet Union was particularly ill-suited due to its economic vulnerabilities to engage in such a wide contest, which became effectively a conflict of attrition. Unsurprisingly, the side with resources to spare prevailed, because while the costs might have been heavy for the US, they proved ruinous for the Soviet Union. It should also be noted that this secondary result was largely unanticipated.[52] Containment thus ultimately worked because the conception behind it was fundamentally sound: as long as the US stayed ahead of the Soviet Union, and the latter was prevented from making critical gains, the status quo could not be overturned.

Notes

1 Baugh, "Great Britain's 'Blue Water' Policy." Semi-detachment was originally an ironic moniker that was affixed on British foreign policy by the French diplomat and journalist Francis de Pressensé in *Le Temps* of 1896. The term was then employed by Howard to sum up the British position toward conditions warranting war. Howard, *Britain and the Casus Belli*, 168, 164–73.
2 The classical expression of this view was provided by US Admiral Alfred Thayer Mahan. For Mahan, certain geopolitical elements, primarily geographical position, allowing a state not to have to mount a defense by land; as well as a favorable physical conformation, such as good harbors and a long defensible coastline, predisposed it toward developing its sea power. Mahan, *Influence of Sea Power*, 29–57.
3 Bough, "Blue Water Policy," 46–8; Bourne, *Foreign Policy of Victorian England*, 7–8.
4 For the naval/ blue water strategy see Corbett, *Some Principles of Maritime Strategy*; Liddell Hart, *British Way of Warfare*; Seton-Watson, *Britain in Europe*, 36–7; also see Milevski, *Evolution of Grand Strategic Thought*, chap. 2.
5 It should be kept in mind that, in the case of a major crisis in the core, relying solely on naval power to help out allies by mounting diversions would be of only limited use. The reason is that the enemy could annihilate the ally, before turning its sights to the naval power, as Napoleon did. A complementary continental expeditionary force is therefore needed to deal with such situations, a reality that has been clear to British decision-makers from the era of Elizabeth I onward. Howard, "British Way in Warfare"; Bough, "Blue Water Policy," 37.
6 Kennedy, "Mahan vs. Mackinder," 43–85, 48–55; Kennedy, *Rise and Fall of British Naval Mastery*, 4, 6, 181–3; Bourne, *Foreign Policy of Victorian England*, 6; Mahan, *Influence of Sea Power*, 27–8, 55–7, 82–3; Friedberg, *Weary Titan*, 219. Britain could mobilize just as many soldiers as the other great powers: in the Crimean War it fielded over half a million troops, on par with France and Austria, and ahead of Prussia.
7 Mearsheimer, *Tragedy*, 261–4, chap. 7; Seton-Watson, *Britain in Europe*, 35–6; Layne, "From Preponderance to Offshore Balancing."

8 Ferguson, *Empire*, 240; Brendon, *Decline and Fall*, 95; Weinberg, *Manifest Destiny*.
9 Levy, "What Do Great Powers Balance Against," 41–4; Levy, "Power Transition and the Rise of China," 11–33, 20–5.
10 Galbraith, 'The "Turbulent Frontier."'
11 Howard, *Britain and the Cassus Belli*, 26.
12 Schelling, *Arms and Influence*, 124.
13 Leffler, *Preponderance of Power*, 19, 15–19; Haas, "Balance of Power"; Levy, "Balances and Balancing," 129–53, 147; Claude, *Power and International Relations*, 41–3.
14 Jervis, *Perception and Misperception*, 58–113; Snyder, *Deterrence and Defense*, 3–40; Schelling, *Arms and Influence*, chaps. 1–2; George and Smoke, *Deterrence in American Foreign Policy*, 59–83; Freedman, *Deterrence*, 8, 26–42.
15 Treisman, "Rational Appeasement"; Friedberg, *Weary Titan*.
16 Jervis, "Domino Beliefs and Strategic Behavior," 20–50; Ravenal, "Counterforce and Alliance," 28; Schelling, *Arms and Influence*, 55–9.
17 Essentially, this was the difference between strongpoint defense (strategic and selective) and perimeter defense (psychological and global). Gaddis, *Strategies of Containment*, 56–60, 89–90; Lippman, "Cold War"; Freedman, *Deterrence*, 52–4.
18 Jervis, *Perception and Misperception*, chap. 3, esp. 76–8.
19 "Memorandum by Mr. Crowe," 407.
20 Kennedy, *Rise and Fall of Great Powers*, 148–9, 154–5, 200–3; Maddison, "Statistics on World Population, GDP and Per Capita GDP"; Kennedy, *Rise and Fall of British Naval Mastery*; Woodruff, *Impact of Western Man*, 314–23.
21 Before 1815, Britain could be described as a rising power competing with France in the world system. Furthermore, up to the 1830s, Britain preserved a vulnerable connection on the continent: the electorate of Hanover, effectively a part of the German world.
22 Kennedy, *Rise and Fall of British Naval Mastery*, 177.
23 Bourne, *Foreign Policy of Victorian England*, 13; Temperley, *Foreign Policy of Canning*, 42–7, 64–8.
24 Portugal is included under this heading too, since it represented a "chokepoint" in the Mediterranean.
25 Ibid., 81–5, 115–17, 133–7, 140–2, 154–5; Bourne, *Foreign Policy of Victorian England*, 14–8.
26 Beloff, *Imperial Sunset*, 37.
27 Mahajan, *British Foreign Policy*, 2–6, 202–3; Friedberg, *Weary Titan*, 220.
28 Kennedy, *Rise and Fall of British Naval Mastery*, 182–3.
29 Ibid., 183–5, 201; Friedberg, *Weary Titan*, 147–8, 144–52.
30 By the end of the century, Britain and Russia were locked in a major rivalry, the so-called Great Game, originally fought over who would control Afghanistan, but later on extended to the Ottoman Empire, Persia, and China. Ingram, "Great Britain and Russia", 269–93.
31 Thornton, *Imperial Idea and Its Enemies*, 44; Howard, *Continental Commitment*, 67; Kennedy, *Rise and Fall of British Naval Mastery*, 183.
32 Farwell, *Queen Victoria's Little Wars*, 1; David Saul, *Victoria's Wars*.
33 "Extract from Palmerston's Reply to His Critics in the House of Commons, 1 March, 1848," in Bourne, *Foreign Policy of Victorian England*, 291–3, 293.
34 Webster, *Foreign Policy of Palmerston*, vol. I, chap. 2.
35 Ibid., 278–89, 333–41; Bourne, *Foreign Policy of Victorian Britain*, 27–9, 34–7; Anderson, *Eastern Question*, 88–93, 90–1.
36 Presently these are the provisions of the Montreux Convention; "Convention Regarding the Regime of the Straits, Signed at Montreux, July 20, 1936," at http://sam.baskent.e du.tr/belge/Montreux_ENG.pdf; Webster, *Foreign Policy of Palmerston*, vol. II, chap, 8; Anderson, *Eastern Question*, 97–107.

37 Kennedy, *Rise and Fall of British Naval Mastery*, 201; Millman, *British Foreign Policy*.

38 In 1914, Britain did not have an alliance with France and Russia as Germany had with Austria, or France had with Russia, but merely an entente. It had joint plans with France to intervene, but no firm commitment to do so. No one knew for certain whether Britain would march to defend France: driven by wishful thinking, Imperial Germany allowed itself to believe it would not. The issue was finally decided by Germany, who violated Belgian neutrality when attacking France. Anderson, *Eastern Question*, chap. 5; Albertini, *Origins of the War of 1914*, vol. II.

39 Leffler, "Emergence of an American Grand Strategy," 67–8, 74–5; Leffler, *Preponderance of Power*, 3–10; Kennedy, *Rise and Fall of Great Powers*, 357–8, 369; Dallek, *Franklin Delano Roosevelt*, 389–90; Gaddis, *United States and Origins of the Cold War*, 25–7.

40 Leffler, *Preponderance of Power*, 12–13, 49–54; Norman Naimark, "Sovietization of Eastern Europe," 175–97; Kuniholm, *Origins of the Cold War in the Near East*.

41 Gaddis, *Long Peace*, 48–9, 51, 58; Isaacson, *Wise Men*, 241–3, 284–5, 351; Dueck, *Reluctant Crusaders*, 86–8; Trachtenberg, *Constructed Peace*, 15–20, 31–3, 34–41. It should be noted that Kennan had initially supported a sphere of influence agreement.

42 Several scholars argue that rather than being the result of Kennan's grand design, containment was arrived at through gradual adjustment, in other words, through learning, trial, and error. Popescu, *Emergent Strategy and Grand Strategy*; Trachtenberg, "Making Grand Strategy"; also see Miscamble, *George F. Kennan*. However, when mentioning earlier US actions to resist the Soviet Union, Trachtenberg does not prove that this amounted to a coordinated, organized stance of resistance, as advised by Kennan, so these would constitute isolated efforts rather than a grand strategy. Meanwhile, Popescu argues that the version of containment that eventually got implemented did not follow Kennan's prescriptions, which is true enough, but does not prove that the substance (the fundamentals of goals, tactics of deterrence and keeping ahead of the Soviet Union, and identification of challenges) was not the same. As Kennan's grand design had to be "sold" to the American public, it is not surprising that it would have suffered changes in terms of specifics, such as rhetoric, emphasis, timing, and extent of application. But according to Popescu's own criteria, if the policy is congruent with the grand design, it should be interpreted as a result of grand, not emergent, strategy.

43 "George Kennan's 'Long Telegram,'" February 22, 1946, at http://digitalarchive.wil soncenter.org/document/116178, 1–5; X, "Sources of Soviet Conduct," 570–3; Gaddis, *Strategies of Containment*, 22–52; Gaddis, *George F. Kennan*, 218–22, 228–30, 259–62. The available evidence from the Soviet side indicates that Kennan was not far off the mark. Stalin did not want a war, but thought that one was inevitable. While he advised caution for communists in Western Europe, he was banking both on driving wedges between Britain and the US in the Third World, eventually driving London into the Soviet embrace, and on exploiting opportunities to expand its influence alongside the Soviet borders. Zubok and Pleshakov, *Inside the Kremlin's Cold War*.

44 X, "Sources of Soviet Conduct," 574–5; "Long Telegram," 5–7; Gaddis, *George Kennan*, 240; Leffler, "Emergence of an American Grand Strategy," 77–8.

45 X, "Sources of Soviet Conduct," 575, 580–2; Gaddis, *Strategies of Containment*, 35–49, 28–9; Gaddis, *George Kennan*, 241; Leffler, "Emergence of an American Grand Strategy," 87.

46 "President Harry Truman's Address Before a Joint Session of Congress, March 12, 1947," at https://avalon.law.yale.edu/20th_century/trudoc.asp; Isaacson, *Wise Men*, 387–402; Leffler, *Preponderance of Power*, 142–6.

47 Isaacson, *Wise Men*, 406–16; Gaddis, *George Kennan*, 241, 264–70; Leffler, *Preponderance of Power*, 157–64.

48 Isaacson, *Wise Men*, 395, 475–8, 480–1, 486–90, 495–6.
49 "April 14, 1950 National Security Council Report, NSC 68 'United States Objectives and Programs for National Security,'" at https://digitalarchive.wilsoncenter.org/document/116191.pdf; Gaddis, *Strategies of Containment*, 88–91, 94–8, 105–6, 115; Christensen, *Useful Adversaries*, 50, 123–8; Leffler, *Preponderance of Power*, 355–60; Isaacson, *Wise Men*, 499–52.
50 The only effort at roll back, the ill-advised attempt to conquer North Korea in the aftermath of the defeat of the North Korean attack on the South, floundered due to the military intervention of China. As the Korean War wound down to a draw, US decision-makers scaled back their ambitions to containment: Communist existing gains were allowed to stand, but new gains were to be resisted.
51 Isaacson, *Wise Men*, 504, 507–9, 526–7; Gaddis, *Strategies of Containment*, 106–15; Leffler, *Preponderance of Power*, 366–7; Stueck, *Korean War*; Yuen Foong Kong, *Analogies at War*, 23; Paige, *Korean Decision*, 136, 144–5, 169–73, 174–9.
52 Westad, *Global Cold War*; Leffler and Westad, *Cambridge History of the Cold War*, vols. 2–3; Kinzer, *Overthrow*; George and Smoke, *Deterrence in American Foreign Policy*; Allison and Zelikow, *Essence of Decision*; Wohlforth, *Elusive Balance*.

Bibliography

Albertini, Luigi. *The Origins of the War of 1914*, vol. II. London: Oxford University Press, 1953.

Allison, Graham and Philip Zelikow. *Essence of Decision: Explaining the Cuban Missile Crisis*. New York: Longman, 1999.

Anderson, MS. *The Eastern Question, 1774–1923*. New York: St. Martin's Press, 1966.

Baugh, Daniel. "Great Britain's 'Blue Water' Policy." *International Historical Review* 10 (February 1988): 33–58.

Beloff, Max. *Imperial Sunset*. New York: Knopf, 1970.

Bourne, Kenneth. *The Foreign Policy of Victorian England, 1830–1902*. Oxford: Clarendon Press, 1970.

Brendon, Piers. *The Decline and Fall of the British Empire, 1781–1997*. London: Jonathan Cape, 2007.

Christensen, Thomas. *Useful Adversaries: Grand Strategy, Domestic Mobilization, and Sino-American Conflict, 1947–1958*. Princeton: Princeton University Press, 1996.

Claude, Inis. *Power and International Relations*. New York: Random House, 1962.

Corbett, Julian. *Some Principles of Maritime Strategy*. New York: Longmans, 1918.

Crowe, Eyre. "Memorandum by Mr. Crowe." In *British Documents on the Origins of the War, 1898–1914*, vol. VI, edited by GP Gooch and Harold Temperley. London: His Majesty's Stationery Office, 1928.

Dallek, Robert. *Franklin Delano Roosevelt and American Foreign Policy, 1932–1945*. New York: Oxford University Press, 1995.

Dueck, Colin. *Reluctant Crusaders: Power, Culture, and Change in American Grand Strategy*. Princeton: Princeton University Press, 2006.

Farwell, Byron. *Queen Victoria's Little Wars*. New York: Norton, 1985.

Ferguson, Niall. *Empire: How Britain Made the Modern World*. London: Penguin, 2003.

Freedman, Lawrence. *Deterrence*. Malden: Polity, 2004.

Friedberg, Aaron. *Weary Titan: Britain and the Experience of Relative Decline, 1895–1905*. Princeton: Princeton University Press, 1988.

Gaddis, John Lewis. *George F. Kennan: An American Life*. New York: Penguin Press, 2011.

Gaddis, John Lewis. *Strategies of Containment: A Critical Appraisal of American National Security Policy During the Cold War*. New York: Oxford University Press, 2005.

Gaddis, John Lewis. *The Long Peace: Inquiries into the History of the Cold War*. New York: Oxford University Press, 1987.

Gaddis, John Lewis. *The United States and the Origins of the Cold War, 1941–1947*. New York: Columbia University Press, 1972.

Galbraith, John. "The 'Turbulent Frontier' as a Factor in British Expansion." *Comparative Studies in Society and History* 2, no. 2 (1960): 150–68.

George, Alexander and Richard Smoke. *Deterrence in American Foreign Policy*. New York: Columbia University Press, 1974.

Haas, Ernst. "The Balance of Power: Prescription, Concept, or Propaganda?" *World Politics* 5 (July 1953): 442–77.

Howard, Christopher. *Britain and the Casus Belli, 1822–1902: A Study of Britain's International Position from Canning to Salisbury*. London: Athlone Press, 1974.

Howard, Michael. "The British Way in Warfare: A Reappraisal." In *The Causes of War and Other Essays*, edited by Michael Howard. Cambridge: Harvard University Press, 1984.

Howard, Michael. *The Continental Commitment: The Dilemma of British Defense Policy in the Era of the World Wars*. London: T. Smith, 1972.

Ingram, Edward. "Great Britain and Russia." In *Great Power Rivalries*, edited by William Thompson. Columbia: University of South Carolina Press, 1999.

Isaacson, Walter. *The Wise Men: Six Friends and the World They Made*. New York: Simon & Schuster, 1986.

Jervis, Robert. "Domino Beliefs and Strategic Behavior." In *Dominoes and Bandwagons: Strategic Beliefs and Great Power Competition in the Eurasian Rimland*, edited by Robert Jervis and Jack Snyder. New York: Oxford University Press, 1991.

Jervis, Robert. *Perception and Misperception in International Politics*. Princeton: Princeton University Press, 1976.

Kennan, George (X). "The Sources of Soviet Conduct." *Foreign Affairs* 25 (July 1947): 566–82.

Kennedy, Paul. "Mahan vs. Mackinder: Two Interpretations of British Sea Power." In *Strategy and Diplomacy, 1870–1945*, edited by Paul Kennedy. London: Harper Collins, 1989.

Kennedy, Paul. *The Rise and Fall of British Naval Mastery*. London: Fontana Press, 1991.

Kennedy, Paul. *The Rise and Fall of Great Powers: Economic Change and Military Contest From 1500 to 2000*. New York: Random House, 1987.

Kinzer, Stephen. *Overthrow: America's Century of Regime Change from Hawaii to Iraq*. New York: Times Books, 2006.

Kong, Yuen Foong. *Analogies at War: Korea, Munich, Dien Bien Phu and the Vietnam Decisions of 1965*. Princeton: Princeton University Press, 1992.

Kuniholm, Bruce. *The Origins of the Cold War in the Near East: Great Power Conflict and Diplomacy in Iran, Turkey, and Greece*. Princeton: Princeton University Press, 1980.

Layne, Christopher. "From Preponderance to Offshore Balancing: America's Future Grand Strategy." *International Security* 22 (Summer 1997): 86–124.

Leffler, Melvyn. *A Preponderance of Power: National Security, the Truman Administration, and the Cold War*. Stanford: Stanford University Press, 1992.

Leffler, Melvyn. "The Emergence of an American Grand Strategy, 1945–1952." In *The Cambridge History of the Cold War, Vol. I*, edited by Melvyn Leffler and Odde Arne Westad. New York: Cambridge University Press, 2010.

Levy, Jack. "Balances and Balancing: Concepts, Propositions, and Research Design." In *Realism and the Balance of Power: A New Debate*, edited by John Vasquez and Colin Elman. Upper Saddle River: Prentice-Hall, 2003.

Levy, Jack. "Power Transition and the Rise of China." In *China's Ascent: Power, Security, and the Future of International Politics*, edited by Robert Ross and Zhu Feng. New York: Cornell University Press, 2008.

Levy, Jack. "What Do Great Powers Balance Against?" In *Balance of Power: Theory and Practice in the 21st Century*, edited by TV Paul, James Wirtz, and Michel Fortmann. Stanford: Stanford University Press, 2004.

Liddell Hart, BH. *The British Way of Warfare*. London: Faber and Faber, 1932.

Lippman, Walter. "The Cold War." *Foreign Affairs* 65 (Spring 1987): 869–84.

Mahajan, Sneh. *British Foreign Policy, 1874–1914: The Role of India*. New York: Routledge, 2002.

Mahan, Alfred T. *The Influence of Sea Power Upon World History 1660–1783*. Cambridge: Cambridge University Press, 2010.

Mearsheimer, John. *Tragedy of Great Power Politics*. New York: W.W. Norton, 2001.

Milevski, Lukas. *The Evolution of Grand Strategic Thought*. Oxford: Oxford University Press, 2016.

Millman, Richard. *British Foreign Policy and the Coming of the Franco-Prussian War*. Oxford: Clarendon, 1965.

Miscamble, Wilson. *George F. Kennan and the Making of American Foreign Policy, 1947–1950*. Princeton: Princeton University Press, 1992.

Naimark, Norman. "The Sovietization of Eastern Europe, 1944–1953." In *Cambridge History of the Cold War*, edited by Melvyn Leffler and Odde Arne Westad. New York: Cambridge University Press, 2010.

Paige, Glenn. *The Korean Decision, June 24–30, 1950*. New York: Free Press, 1968.

Popescu, Ionut. *Emergent Strategy and Grand Strategy: How American Presidents Succeed in Foreign Policy*. Baltimore: Johns Hopkins University Press, 2017.

Ravenal, Earl. "Counterforce and Alliance: The Ultimate Connection." *International Security* 6 (Spring 1982): 26–43.

Saul, David. *Victoria's Wars: The Rise of Empire*. London: Penguin, 2009.

Schelling, Thomas. *Arms and Influence*. New Haven: Yale University Press, 1966.

Seton-Watson, RW. *Britain in Europe: A Survey of Foreign Policy*. New York: Macmillan, 1937.

Snyder, Glenn. *Deterrence and Defense*. Princeton: Princeton University Press, 1961.

Stueck, William. *The Korean War: An International History*. Princeton: Princeton University Press, 1995.

Temperley, Harold. *The Foreign Policy of Canning, 1822–1827: England, the Neo-Holy Alliance, and the New World*. London: Frank Cass, 1966.

Thornton, AP. *The Imperial Idea and Its Enemies*. London: Macmillan, 1959.

Trachtenberg, Marc. *A Constructed Peace: The Making of the European Settlement, 1945–1963*. Princeton: Princeton University Press, 1999.

Trachtenberg, Marc. "Making Grand Strategy: The Early Cold War Experience in Retrospect." *SAIS Review* 19 (Winter 1999): 33–40.

Treisman, Daniel. "Rational Appeasement." *International Organization* 58 (Spring 2004): 345–73.

Webster, CK. *The Foreign Policy of Palmerston, 1830–1841*. London: G. Bell & Sons, 1951, vol. I.

Webster, CK. *The Foreign Policy of Palmerston, 1830–1841*. London: G. Bell & Sons, 1951, vol. II.

Weinberg, Albert. *Manifest Destiny: A Study of Nationalist Expansionism in American History*. Baltimore: Johns Hopkins Press, 1962.

Westad, Odd Arne. *The Global Cold War: Third World Interventions and the Making of Our Times*. Cambridge: Cambridge University Press, 2005.

Wohlforth, William. *The Elusive Balance: Power and Perceptions During the Cold War*. Ithaca: Cornell University Press, 1993.

Woodruff, William. *Impact of Western Man: A Study of Europe's Role in the World Economy, 1750–1960*. New York: St. Martin's Press, 1967.

Zubok, Vladimir and Constantin Pleshakov. *Inside the Kremlin's Cold War: From Stalin to Khrushchev*. Cambridge: Harvard University Press, 1996.

6 The grand strategies of declining powers

This chapter explores the grand strategies available to declining powers, that is to say those powers that experience a consistent reduction of power and status relative to the other members of the club. The grand strategies considered comprise appeasement and lying in wait. As in previous chapters, the chapter begins by mentioning the key points of each grand strategy before offering historical illustrations of the grand strategies.

Appeasement

Appeasement is a grand strategy whose objective is to avoid, postpone, or minimize the risk of war. First, this objective has to do with dodging the prohibitive costs of war. This is the case when the great power is weaker than its main opponent, and cannot compensate its weakness through alliances, given that either it faces a stronger coalition than the one it is able to put together itself, or that it is isolated among enemies. When war is a losing prospect, there is no reason to undertake it, and a powerful incentive to steer clear of situations that can provoke it. Second, by appeasing, the great power may get breathing room to recover its strength, find new allies, or hope that its enemy will make a mistake. Third, appeasement is beneficial for triage purposes. If the great power is overextended, and faces too many challenges at once, appeasing secondary opponents allows it to redistribute resources in order to resist the chief challenger. Fourth, appeasement may deflect a would-be aggressor against a third party, and in so doing, enmesh it in a war of attrition that would end up weakening it.[1]

Appeasement is pursued through several tactics implemented simultaneously. In the first place, there is retrenchment, understood as the unilateral curtailing by a great power of its foreign commitments. The great power may withdraw its military, give up particularly exposed dependencies or spheres of influence, exit risky alliances, and stop providing military, economic, and diplomatic support for clients, both governments and non-state actors, when such support is deemed too costly.[2] Appeasement is pursued also through passive appeasement, meaning a non-provocative stance. If a potential opponent is looking for a pretext to initiate war, it would be foolish to provide it with one. Thus, it is smarter to keep a low profile by purposefully keeping away from bellicose rhetoric or measures that could be interpreted or cast as hostile and, as such, offer the justification the enemy needs to start fighting.[3] Another tactic is active appeasement or conciliation, that is reducing and preferably permanently removing antagonism by giving a sympathetic hearing to the opponent's grievances, and by satisfying them through negotiations. Conciliation is not the same thing as détente, understood as the relaxation, but not the complete elimination, of tensions between two powers that otherwise remain locked in conflict, for instance the US and the Soviet Union in the 1970s. In détente, the powers agree on compromise terms that do not confer a sizable advantage to either side. Meanwhile, conciliation seeks the peaceful resolution of the root issues causing conflict on terms largely favorable to the party pressing its demands or grievances.[4] But perhaps the single most important tactic of appeasement consists in offering unreciprocated concessions. This makes appeasement different from engagement, which works through the extension of mutual concessions, hence by both sides. By contrast, in appeasement, one side provides the other with concessions without reciprocity, in return for a promise not to take any action that would risk bringing about war. This may refer to either the challenger refraining from direct attack, or from attacking an ally, dependency, or client, which then the appeaser would have to defend.

These individual tactics often figure as part and parcel of other grand strategies. For instance, Louis XIV employed unreciprocated concessions in his grand strategy of primacy, successfully detaching some of his enemies, such as Savoy and the German princes, from the coalition opposing him. Great Britain alternated conciliation with intervention in Europe as part of its grand strategy of semi-detachment, for instance by staying away from the wars of unification of Italy and Germany. Deng Xiaoping's China employed a low-key policy in its grand strategy of reassurance and temporization, so as not to cause an accidental escalation. Even US containment resorted to conciliation, especially in its minimalist version, which prioritized protecting areas valuable in terms of military and industrial resources, but abstained from resisting the advance of communism in other quarters. That is to say appeasement as a strategy does not designate any one of the above tactics, but rather their use in conjunction with each other for the purpose of warding off war.

There are at least two variants of appeasement as a grand strategy. One is a hybrid grand strategy, mixing appeasement and containment. A great power, especially a dominant one, faces up to a multitude of challengers in various regions of the world. Although it is still powerful enough to defeat piecemeal any of the challengers, it no longer has sufficient strength to fend off all, or even several of them, if attacked at the same time. Hence, the great power resorts to prioritizing its resources, transferring them from less important regions and toward the key strategic theater, sacrificing the first in order to obtain a decisive advantage in the latter. Essentially, appeasement as half of the grand strategy is a way to bring goals, means, and challenges back into synch.[5] In this variant of appeasement, the other half of the grand strategy is the containment of the chief opponent. Appeasement is hence not followed all the way through—malleability in one arena allows for rigid resistance in another.[6] The other variant of appeasement is a wholesale approach, in which a great power seeks to stay out of conflict all around while buying time, with the expectation that it may be able to mount a more successful resistance later on from an improved position of power. At that time, it may either decide to go to war or confront the opponent with the tables now turned.[7]

Appeasement has been famously compared by Winston Churchill to a man feeding a crocodile in the hope that the beast will eat him last. Appeasement is regularly taken to task for rewarding the opponent for engaging in threatening or aggressive behavior, which encourages it to press even harder, in the hope of a constant or bigger payoff. Therefore, appeasement opens the door to extortion, in which a great power extracts periodic payment (tribute), just as in a protection scheme run by a gangster in a given neighborhood. As Machiavelli put it:

> for if you yield through fear and to escape war, the chances are that you do not escape it; since he to whom, out of manifest cowardice you make this concession, will not rest content, but will endeavor to wring further concessions from you, and making less account of you, will only be the more kindled against you.

Appeasement also is blamed for signaling weakness: if one aggressor is paid, this constitutes an incentive for others to follow suit. Thus, appeasement lays waste to the credibility of one's deterrent threats: even if the defender, having given in on one issue, now has the utmost resolve to stand up and fight, who would believe it? Finally, from a cost-benefit perspective, it can be argued that appeasement is more expensive than containment, which relies on deterrence, since it tries to buy off the opponent with concessions in terms of territory, wealth, influence, or status in exchange for being left in peace. By contrast, all that is needed in deterrence to enjoy peace is to stand up to the aggressor from the get-go. Faced with determined resistance, it is likely to back off, meaning that the defender does not have to pay it anything.[8]

Contemporary scholarship, nevertheless, has reassessed the virtues of appeasement. The grand strategy is nowhere as bad as it has been made up to be. For a weaker party, there is no advantage in starting a war that it would lose. This would be like seeing a particularly brave lamb charge an incoming wolf—there are few chances the predator would back away. As Thucydides put it in the Melian debate, in relations between states, the strong do what they can, while the weak suffer what they must. Being the weaker side is not enjoyable or desirable, but it is a reality that cannot just be wished away, and which imposes a different behavior from that of an actor enjoying the upper hand. Even being eaten last has its advantages. In Homer's *Odyssey*, Ulysses and his companions are captured by the cyclops Polyphemus. Knowing they cannot overcome this stronger foe, Ulysses resorts to trickery: he plies the cyclops with wine, in return for the promise of being eaten last. He thus avoids the fate of some of his companions who immediately end up in the cyclops' belly, and after Polyphemus falls into a drunken sleep, he then proceeds to blind him with a fire-hardened stake. For an appeaser, time is the most precious resource: it makes sense to fight later, rather than fight now from a position of weakness. The appeaser is simply willing to pay for postponing the moment of confrontation. Furthermore, appeasement may be a necessity if one confronts not only one, but multiple challengers—successful resistance may not be possible unless the great power avoids war with at least some of these would-be enemies. This is acknowledged by Machiavelli, who admits that his injunction applies only to cases

> where you have a single adversary only; but should you have several, it will always be a prudent course, even after war has been declared, to restore to some one of their number something you have of his, so as to regain his friendship and detach him from the others who have leagued themselves against you.

Besides, research in the last quarter-century has called into question whether there is an actual credibility loss for a power backing off in one context, which then leads to a multiplication of extortionist demands from the same challenger or from others. One study found only a single instance out of 16 cases of one party backing off in crises, where its deteriorating reputation then seemed to have had an emboldening effect on the opponent. Further scholarship has cast doubt on the validity of reputation effects, questioning whether a new government necessarily inherits the reputation of its predecessor, whether reputation is affected by recent appeasing behavior or by an older and longer record of firmness, or whether a great power yielding on one issue would yield on an unrelated issue, with reputation being issue-specific. It is also highly debatable whether containment is necessarily cheaper than appeasement. While containment may be cheaper, assuming that deterrence works, it is also the riskier strategy if deterrence fails, for instance when confronting a challenger willing to pay a high penalty for carrying out aggression anyway. In such a scenario, costs are likely to be considerably higher than postponing war through concessions.[9]

Finally, the choice between containment and appeasement is a false one, because each grand strategy works for a different type of power. If the great power enjoys a superior position of strength versus the opponent, being either a dominant or an established power, it is reasonable for it to stand its ground when threatened rather than offer concessions, because it is the one that holds the better cards. If the challenger presses on, it will be the one to face defeat. Even if the defender enjoys less of a power advantage relative to the challenger, which may make a confrontation unpalatable, it may be able to ask for tangible concessions in turn, as in the case of concert or engagement. But for a declining great power, which has been already surpassed by a challenger, and is short of both resources and powerful allies, appeasement may be the best way out of its predicament, because the risk of any additional confrontation under the existing circumstances is unacceptable. Any crisis may spiral out of control and end in a war, which would impose costs that the great power is no longer able to shoulder. Therefore, war would worsen structural decline, instead of helping rectify it, and may provoke irretrievable disaster. To this extent, the appeaser's cardinal rule is not to add to the losses the declining power has already suffered by engaging in reckless confrontation. This makes appeasement the poor man's choice. The main flaw of the grand strategy is that, despite the best efforts of its practitioner, war may be in the end unavoidable, since the initiative of launching it rests entirely with the challenger. If, indeed, going to war is of less value to the challenger than the payoff received, it may hold off attack or turn away toward other targets. But if it were to judge the payoff insufficient, it will press on, ultimately causing the appeaser to decide that, come what may, it has to make a last stand.

Lying in wait

It is not often throughout modern history that a great power experiencing severe decline manages to make a comeback, let alone return even stronger. Examples of successful comebacks are preciously few and limited to established powers: France after its defeats in war in 1870–1871 and 1940, the Soviet Union after World War I, and Germany and Japan after World War II. However, if decline is difficult to reverse, it is not impossible. Lying in wait designates a grand strategy that is tailored to produce a comeback based on three steps: reforms, caution and aloofness, and auctioning. This grand strategy is best illustrated by Stalin's Soviet Union.

The single most important measure to take for a great power wishing to make a comeback is to address the root causes of decline by implementing profound reforms. In terms of structural decline, this entails, on the one hand, belt-tightening by reducing expenses, and, on the other hand, increasing revenues by putting into use previously untapped resources, or by managing the existing ones more efficiently, through importing new economic and military technologies, administrative techniques, fighting doctrines and training, and education of the work force.[10] In terms of political decline, this requires setting up a strong government that is both able to put to rest or suppress old quarrels and to exercise authority efficiently. Consequently, the government is better placed to levy taxes and raise armies without being hampered by domestic hurdles. Again, addressing structural and political decline are interdependent processes: solving one may require progress in the other. A great power may know what needs to be done structurally, but find it impossible to carry out the reforms needed, because it risks social and political upheaval. This was the case of Imperial China under Li Hongzhang and of the Ottoman Empire under the Young Turks. Alternatively, a politically functional great power may reject the need for or postpone carrying out structural reforms, all the while falling further behind its great power rivals. This was the case of nineteenth-century Austria-Hungary or of pre-revolutionary Russia.[11]

However, in the case of a declining power attempting a comeback, unlike in the case of the grand strategy of self-strengthening discussed in Chapter Three, reforms by themselves are not enough. Their effect is to arrest decline, but not necessarily reverse it, which would require scoring a series of gains. The danger at this point is that the declining great power, once somewhat strengthened, engages in frenetic diplomatic and military activity, which then draws the hostility of the other great powers. Therefore, the barely recovered power may nullify all its previous efforts by getting mixed up in conflict before it is fully ready, thus squandering away its hard-won new resources. While victory is prohibitively costly at this stage, defeat may return it to square one. The cautionary example here is Napoleon III's France. The later Napoleon came to power on the promise of reversing the defeat at Waterloo and the Congress of Vienna settlement, and, in so doing, regaining the dominant power and status France had enjoyed during the reign of his uncle. As Napoleon III's domestic political foundation was weak, he engaged in a hasty and chaotic effort to expand in any quarter where expansion could be achieved. While obtaining a number of limited victories, France ended up alienating all the other great powers, which eventually left it diplomatically isolated. The catastrophic defeat at the hands of Bismarck's Prussia then spelled the end of the French efforts to regain its former number one spot.[12]

This is why caution must accompany reforms: the great power has to sit out wars, decline participation in military interventions, and, as much as possible, abstain from drawing attention to itself either through its actions or its rhetoric, especially by airing publicly its grievances with the existing distribution of power and status. As far as the other great powers are concerned, the message that it should convey to them is that it has learned to live with a substantial reduction of its former position in the club and that it is perfectly content with its new circumstances. This is of course pretense—but the great power should not give up the game before it is ready or the opportunity presents itself. As a result of its seeming passivity, the other great powers discount it as an immediate threat, while turning their attention toward the threat posed by each other. While they are busy contending, the declining great power patiently adds to its strength.

The declining great power should not only abstain from making waves, but should also strive to stay out of the existing alliances and alignments. Committing too early to any one of the existing sides means that it may be dragged into a conflict by its ally and in doing so, it would have to give up on its project to continually increase its strength. But an equally important consideration is that premature commitment prevents it from extracting full benefits out of its unaligned stance: it would give up its allegiance for free, instead of capitalizing on it. Indeed, considering that the other great powers have already grouped up into alliance blocs, and, assuming that its reforms have been successful, it would be the only powerful actor that is left that has not chosen a side. Hence, its initial aloofness eventually constrains the alliance blocs to bid for its support. It is only at this stage, similar to running an auction, with its allegiance ultimately going to the highest bidder, that the declining great power can dispense with dissimulation and proclaim its ambitions openly from a position of strength. Effectively, it proposes a quid pro quo in extending its assistance or its neutrality to one of the existing sides in return for recovering its former territorial possessions, sphere of influence, or position in the club. As such, the declining power does not act so as to create its own opportunity for a comeback, but waits for the other powers to create it for its benefit, and it is only at this stage that it moves to seize it. Hence, the great advantage of lying in wait: it minimizes the opposition the great power has to face in making a come-back, and the cost it has to pay.

At first glance, lying in wait has much in common with what International Relations scholars refer to as buckpassing. Buckpassing refers to a state that, instead of balancing against the threat of another state that is seeking hegemony, prefers to sit the conflict out and let other great powers resist it in its stead, in so doing passing them the buck. The result is that as the wannabe hegemon and the opposing coalition exhaust themselves fighting each other, the buckpassing great power rides free: it reaps the benefit of security it does not have to pay for.[13] Ancient Chinese strategy put it more colorfully: sitting on top of the mountain and watching the tigers fight.

Yet, lying in wait differs from buckpassing. In lying in wait the great power eventually must commit itself to taking part in the on-going dispute, instead of remaining permanently detached. The reason for choosing a side in the end is that the declining power runs the risk of the other great powers striking a deal behind its back and then proceeding to move jointly against it. This is what befell pre-World War II Poland, which stubbornly insisted on staying neutral, instead of joining with either the Soviet Union or Nazi Germany, and, as a result, suffered depredation at the hands of both. Even more ominously, one of the existing sides may play the very same game that the lying-in-wait power plays, by deflecting the aggression of its erstwhile enemy against it, while being the one to sit the conflict out. For this reason, to avoid being cheated itself, instead of being the cheater, it is in the best interest of the declining power to strike a deal with one of the two sides. The whole skill is to succeed in doing so without conveying the impres-sion of being too eager to conclude an alliance, or being willing to do so at any price, but rather making sure that the prospective partner pays up for its services. Once substantial concessions are being offered, there is no reason for the power to maintain its aloofness, as is the case in buckpassing.

Lying in wait is also different from the rising power grand strategies of tertius gaudens/divide and conquer discussed in Chapter Three. Tertius gaudens is actu-ally based on the pursuit of multiple crisscrossing alignments, as exemplified by Bismarck's alliance system. The rising power practicing this grand strategy is not at all aloof, but uses some of its partners to neutralize others, leaving it with an open path to advancement. Lying in wait, by contrast, is about delaying as much as possible committing to any side, and eschewing any risky policy of advance-ment that may backfire on it.

Lying in wait has two important limitations. First, it is a grand strategy requir-ing a good deal of patience, possibly even stretching across multiple generations. The declining great power needs time to implement reforms, amass new capabili-ties, all-the-while waiting for the right moment to make its move. This process of waiting and lying low may be very long in the making, assuming that the right moment arrives at all, which might make it infeasible due to domestic pressures to deliver speedy results. Second, the other danger of lying in wait is that the prac-titioner might be too smart for its own good: having decided to sit out a conflict in the belief that the resulting fight will bleed white the other sides, which would make them compete for its allegiance, it gambles on a drawn-out confrontation between them. But if one side were to emerge as a clear winner in a fast contest, the ability of the declining power to exert concessions from it would be severely curtailed. This is again what happened to Napoleon III, who counted on Prussia and Austria to fight each other to a standstill, only to bear witness to the military triumph of Prussia within six weeks and have Berlin emerge as a much more formidable rival.

Appeasement: Great Britain and decline

The case of Great Britain is a tale of two instances of appeasement. In the first instance, at the cusp of the twentieth century, Britain opted for a hybrid strategy of containment and appeasement. In the second instance, extending from the aftermath of World War I and up to World War II, Britain put in place a comprehensive strategy of appeasement, aimed at postponing and avoiding any major conflict. In the former instance, Britain met with resounding success in rebalancing its commitments and standing up to its main opponent, getting a lease of life in its tenure as dominant state. But the latter instance has been widely seen as an unmitigated disaster, due to a well-entrenched perception that Britain's appeasement by the so-called *Guilty Men* had abetted Hitler in his aggressive plans.[14] What accounts for these contrasting outcomes? This case study suggests that there is nothing wrong with appeasement as a grand strategy, but rather that problems showed up in how the grand strategy was developed and implemented the second time around.

In the first interval, stretching from the 1890s to World War I, London was not actually a declining power. Admittedly, it had fallen behind both the US and Germany in terms of industrial production. By 1900, Britain's share of global manufacturing output had fallen to 18.5%, while Germany's had increased to 13.2%, and the US had already surpassed it with 23.6%. By 1913, Britain had dropped even further behind with an output of only 13.6% versus 32% and 14.8% for Washington and Berlin, respectively. But this setback had taken place in a single dimension, and its full extent was not fully realized, as there were no macroeconomic indicators at the time that would have allowed measuring just how much Britain was lagging. Faced with this lack of a smoking gun, Prime Minister Arthur Balfour was musing in 1903 whether Britain was actually worse off:

> is our position going to worsen relatively to other nation, or even to worsen absolutely? I see no satisfactory symptoms ... judged by all the available tests, both the total wealth and the diffused well-being of the country [are] greater than they had ever been.

In effect, in almost all the other important dimensions, London remained ahead of everyone else. In trade, it enjoyed a 20% share of world trade compared to second-place Germany's 13% in 1900; it deployed a navy larger than the next two great powers combined; and in colonial matters, it was the owner of by far the larger and most rapidly expanding empire. Between 1876 and 1915 the British Empire grew by 4 million square kilometers, compared to an increase of 3.5 million for France and 1 million for Germany. Therefore, to quote Paul Kennedy, due to the "combination of financial resources, productive capacity, imperial possessions, and naval strength … [Britain] was still probably the 'number-one' world power."[15]

But staying on top had never seemed so uncomfortable. As Lord Thomas Sanderson, Permanent Under-Secretary at the Foreign Office, put it, Britain "must appear in the light of some huge giant sprawling over the globe, with gouty fingers and toes stretching in every direction, which cannot be approached without eliciting a scream." To sum up its dilemma, Britain was confronting at the same time five other great powers in a contest spread over five different regions. In the Western Hemisphere, it contended with a rising US; in Africa, it confronted both France, especially over Egypt, and Germany, especially over South Africa; in the Mediterranean it stood up to the Dual Alliance of France and Russia; in Far East Asia, particularly China, it faced both Russia and the newcomer Japan; and, finally, in the Great Game, it opposed Russia in Persia, Afghanistan, and Central Asia. Since all of these competitors were aggressively engaging in the building of warships and the pursuit of colonies and overseas bases, Britain could no longer count on semi-detachment, keeping its involvement in the core limited, so as to free sufficient resources for it to enjoy unrivaled supremacy in the periphery. Even though Britain could have probably emerged the victor against any single opponent in any of these separate areas, it knew that the minute it engaged in one such contest, it would have no longer been able to tackle the rest. Simply put, the empire was too big to sustain. As Joseph Chamberlain, Secretary for the Colonies, put it, Britain had become "the weary titan [that] staggers under the too vast orb of its fate."[16]

Accordingly, the solution thought up by British decision-makers was to move from the prioritization of periphery over the core to a substantially different approach, one that placed increased value on some parts of the periphery over other areas also located in the periphery. Thus, Britain was to appease some of its contenders in the regions deemed of lesser importance so as to be able to transfer resources to the defense of critical regions. Considering that British India together with its access routes was paramount to Britain conserving its empire, this meant that the Indian subcontinent and subsequently the Middle East, Central Asia, the Mediterranean, and parts of Africa such as Egypt and South Africa were key regions, while other commitments, more remote or disconnected from India, such as the Western Hemisphere and the Far East, were expendable. Consequently, Britain professed a willingness to compromise with the challengers in the later regions, respectively the US and Japan, so as to better stand up to those in the former, respectively the Dual Alliance of Russia and France.

The need for such a step became stringent in the context of deciding the number of warships Britain needed to build. As seen in Chapter Five, the navy was seen as crucial to Britain's existence as a great power because it enabled global trade and interactions with the colonial empire. As the Admiralty stated: "the

British Empire floats on the British Navy. It is our all in all. Victory at sea, desirable to foreign States is a sine qua non to our continued existence."[17] However, by 1901, it had become clear that the long-standing Two-Power Standard could no longer be maintained if the US kept up its naval build-up, and surpassed the Russian navy as the third naval power in the world after Britain and France. Since Britain's main contender was the Dual Alliance, London would have had to effectively build enough warships to stand up to three, not just the two next-in-line powers. As Lord Selborne, the First Lord of the Admiralty concluded in a ground-breaking memorandum, "it is certain that it would be a hopeless task to attempt to achieve an equality with the three largest navies." Therefore, he made the case for the Two-Power Standard to no longer be calculated numerically, but always in reference to Britain's most likely opponents, i.e., France and Russia.[18] But once there, Selborne and the Admiralty went further, arguing for a hierarchy of regional objectives. As the Admiralty put it,

> Great Britain unaided can hardly be expected to maintain in the West Indies, the Pacific, and in the North American stations, squadrons sufficiently powerful to dominate those of the United States and at the same time to hold the command of the seas in the home waters, the Mediterranean and the Eastern Seas, where it is essential that she should remain predominant.

The same situation was present in the Far East:

> if the British navy were defeated in the Mediterranean and the Channel, the stress of our position would not be alleviated by any superiority in the Chinese seas. If, on the other hand, it were to prove successful in the Mediterranean and the Channel, even serious disasters in the Chinese seas would matter little.

Hence, the next natural step was to compromise on these secondary interests so as to shore up the main interests. In so doing, Selborne was not proposing that Britain cut loose and run away. Britain did not actually give up its possessions in either region, such as Canada or Singapore. Instead, Selborne had in mind appeasement, in which Britain would conclude partnerships or alliances with the emergent powers in the regions it was planning to withdraw its naval forces from, effectively transferring upon them part of the responsibility, and eventually the full responsibility, of defending local British interests against Russia and France. As Selborne saw it, "the case would bear a different aspect were we assured the alliance of Japan ... Great Britain would be under no necessity of adding to the number of battleships on the China station ... our Far Eastern trade and possessions would be secure."[19] Thus, Britain was gaining in two ways. First, it no longer had to spend its own resources to protect secondary regions, and instead was afforded security in those quarters by the US and Japan free of charge. Second, it was able to concentrate its forces where it really mattered, facing the challenger there with superior strength.

This argument eventually won over the British military and civil establishment. In 1903, the British War Office proclaimed that since

> in the event of war with the United States coinciding with a time of uncertain relations between this country and a European Power, the conclusion appears to be unavoidable that the present strength of the H.M.'s navy would not suffice to defend on the high seas the interests of the Empire ... [hence] the contingency of war with the United States should be avoided at all hazards.

In 1905 the Admiralty concurred:

> the more carefully this problem is considered, the more tremendous do the difficulties which would confront Great Britain in a war with the United States appear. It may be hoped that the policy of the British Government will ever be to use all possible means to avoid such a war.

The argument that worked for the US was also applied to Japan. As Sir George Clarke, the chairman of the Committee for Imperial Defense (CID), the highest British security forum put it:

> what is best not to say is that we believe that the idea of opposing the navy of the US in the Caribbean and the North Atlantic close to its bases must be abandoned. ... In years not far distant we shall be quite unable to oppose the navy of Japan in its own waters. It is best to recognize facts but not always to proclaim them from the housetop.[20]

Accordingly, Britain quietly underwent a strategic revolution. Whereas in the early 1890s, London had expressed uneasiness toward the growth in capabilities and ambitions of the US and Japan in the Western Hemisphere and the Far East, it spent the next decade heaping concessions upon both. In relation to the US, Britain was the only great power to support it in its 1898 war against Spain; it endorsed America's terms for the resolution of the Alaskan Panhandle dispute with Canada in 1900; it accepted on two occasions in 1901 revisions favorable to the US concerning the building of the Panama transoceanic canal; and ultimately decided to withdraw altogether its military presence in the Western Hemisphere in 1906. In relation to Japan, Britain was the first power to recognize Tokyo as a sovereign state in 1894, paving the way for the Japanese repudiation of the unequal treaties that had been forced upon it by the European powers. Britain then sided with Japan, rather than with its former protégé China in the Sino-Japanese War, by declining to take part in the intervention of Russia, France, and Germany that forced Tokyo to give up most of its gains from the war. In 1902, Britain shocked the world by concluding an alliance with Japan, London's first alliance in near half a century, which then proved instrumental in deterring France from helping out Russia in the war with Japan in 1904–1905. In 1906, mirroring its actions in the Western Hemisphere, Britain withdrew the bulk of its naval forces from the

Far East. Therefore, these measures were evidence of more than just the use of an isolated tactic, but rather indicated that Britain was relying on a combination of retrenchment, conciliation, and unreciprocated concessions. Britain did not just withdraw militarily; it effectively removed any reasons for clashing with the US and Japan, now unrivalled in their own regions; and it did this by surrendering its dominant position in the Western Hemisphere and the Far East without receiving anything tangible in return other than the promise of being left in peace.[21]

Nonetheless, this promise turned out to have a crucial empowering effect. By 1906, Britain was able to relocate its naval contingents away from the two regions, and, consequently, reached the point where, once again, it enjoyed a lop-sided naval advantage in Europe, the Mediterranean, and the Middle East over a combination of not just France and Russia, but also with Germany thrown in. The ratios post-transfer showed 53 battleships built and 8 building for Britain, 16 built and 6 building for Germany, 20 built and 6 building for France, and a meager 10 surviving ones for Russia in the aftermath of its defeat to Japan. Britain no longer took into account the otherwise impressive number of 24 battleships owned by the US or those of the victorious Japanese navy. As a result, Britain was no longer a weary titan, but once again a dynamic power in the pole position, free to resort to containment, making sure that it conserved its superiority versus its chief competitor. It is primarily due to this regained strength that the Dual Alliance eventually realized that competing against Britain was pointless and decided to negotiate ententes with London. Unlike the arrangements with the US and Japan, the two ententes were distinctly based on mutual concessions. In fact, they were concluded on terms highly favorable to British interests: the recognition of a British sphere of interest in both Egypt by France and in southern Iran and Afghanistan by Russia. When Germany then pressed on its challenge to British naval supremacy from 1908 onwards, Britain had the free hand it needed to be able not just to match it ship for ship, but also to eventually out-build it, by committing to laying two keels for every new battleship launched by Berlin. This would have been a course of action impossible to contemplate only six years beforehand. For all intents and purposes, appeasement had saved the day for Britain.[22]

Britain's initial successful experimentation with appeasement recommended it as an ideal solution for dealing with the country's predicament in the decades following World War I. By that time, Britain's position had considerably worsened, although, paradoxically, a cursory look at a map would have shown its empire at its apex. The Prime Minister of South Africa even referred to it as "the greatest power in the world," and expressed the view that "only unwisdom or unsound policy could rob her of that position." After the annexation of former German colonies and of mandates from former provinces of the Ottoman Empire, the empire stretched in an unbroken line from the Suez to Cape Town, and from Cairo to India. But this impression of strength was belied by structural weakness. In the words of Michael Howard, "Sanderson's 'gouty giant' had become a brontosaurus with huge, vulnerable limbs which the central nervous system had little capacity to protect, direct or control." Britain had not only not given up anything of the huge area it had owned at the start of the century, but had also acquired even more places to defend. Furthermore, imperial defense got far harder and more expensive due to several other simultaneous developments. First, the spread of nationalism in India and the Middle East meant that Britain was on permanent call to police its turbulent possessions against movements seeking independence. Second, Britain had traditionally counted on the resources provided by its empire, especially by India and by the Dominions (Canada, Australia, New Zealand, South Africa, and Ireland). But in 1931, by the Statute of Westminster, Britain recognized the long-standing demand of the Dominions for the status of autonomous communities in the empire. This meant that these countries were admitted as being de facto independent from Britain, therefore, free to control their own military budgets and to pursue their own foreign policy. But even though they were no longer obligated to support Britain and to contribute to its protection, Britain remained, on the other hand, committed to defend the Dominions. As Corelli Barnett sums up the consequences, Britain was "absolutely on her own so far as Europe was concerned, just as if she possessed no empire. Yet at the same time she still retained an unlimited obligation outside Europe to protect the empire. It was hardly a good bargain."[23]

Adding to Britain's woes was that it was supposed to do even more than at the turn of the century while relying on a shrinking economic and military foundation. Like every other great power, Britain had been hit by a double whammy: first the ruinous expenditures of the World War, followed by the Great Depression of 1929. But unlike other powers that could count on superior industrial production to get back on their feet, in the 1930s, Britain's GDP had fallen to fourth place, behind the US, the Soviet Union, and Germany. At the same time, Britain's traditional superiority in trade and finance had been all but annihilated—by the 1930s, its share of world trade fell below 10%, tied up with France and Germany. Meanwhile, with its foreign currency and gold reserves drying up to pay for World War I debts to the US, Britain had turned from the world's foremost creditor into a debtor nation. As a result, and taking into account the absence of manifest external threats in the 1920s, Britain took the disastrous step of adopting the Ten Years' Rule, which stated that "the British Empire will not be engaged in any great war during the next ten years." Hence, Britain had no need for a large military establishment. Its military forces were slashed in the 1920s by as much as 50%, without considering the costs of having to reconstitute them. For the first time in a hundred years, Britain accepted naval parity with the US, and a ratio of tonnage of 5:3 compared to Japan, and 5:1.75 compared to France and Italy. By 1934, the British air force ranked fifth in the world. Britain had never excelled in ground forces, but now was decisively outgunned and outnumbered by the armies of the Soviet Union, France, Germany, and even Italy, which meant that with its own troops tied down by imperial police duties and with no reliable ally, Britain could not have intervened to influence the direction of events on the continent. This time around, unlike the 1900s, Britain was in unmistakable structural decline.[24]

This situation might not have been desperate, were it not for the deteriorating circumstances of international security the 1930s, with the coming to power of revisionist regimes in Italy, Japan, and Germany, their subsequent increase in armaments, and, finally, their pursuit of aggressive expansion in the Mediterranean and Africa, East Asia, and Eastern Europe. To its credit, Britain recognized the problem from the get-go. In 1932, in the face of the Japanese seizure of Manchuria, the previously mentioned CID recommended that the Ten Years' Rule be abandoned. Immediately following Hitler's election as Chancellor of Germany in 1933, it created a subcommittee to assess defense deficiencies, the so-called DRC (Defense Requirements Subcommittee). In its first report, the DRC named Germany as the greatest danger that Britain was facing. However, the subcommittee was painfully aware that, in light of its economic troubles, Britain needed a long time to rearm in order to successfully oppose the challengers. Britain was in no condition to fight even the weakest of the three powers, because had it done so, the other two were bound to take advantage of its being pinned down in fighting to secure large gains in their regions. As the Chiefs of Staff argued,

> a cardinal requirement of our national and imperial security [is] ... to avoid the possible development of a situation in which we might be simultaneously confronted with the hostility ... of Japan in the Far East, Germany in the West, and any power on the main line of communication between the two [Italy.]

Consequently, the DRC recommended that while rearmament was taking place, "it was of the utmost importance that this country should not become involved in war within the next few years." But this was not just a simple matter of deciding to rearm and staying out of trouble while doing so. Britain did not have sufficient resources at its disposal, which meant that it was confronting an impossible choice between economic recovery on the one hand, and rearmament on the other. As the choice for either option at the expense of the other proved unpalatable, Britain attempted to do both through appeasement.[25]

Appeasement in the inter-war period was foremost the design of Neville Chamberlain, the son of Joseph Chamberlain of "weary titan" fame, Chancellor of the Exchequer, and future Prime Minister. Despite much later vilification, Neville Chamberlain was not a naïf optimist, who put his trust in the effectiveness of international rules, the League of Nations, and the natural harmony of relations between states, or a vacillating politician, hesitating to stand up to an aggressor. In fact, he was both a realist, and one of the more iron-handed British political leaders in the twentieth century.[26] As early as 1934, he prognosticated that the quarter from which the next threat would come would be Germany, "the bully of Europe." However, a businessman by training, and as the naturally thrifty Chancellor of the Exchequer, Chamberlain was suspicious of allowing a free-hand to uncontrolled military spending. Germany had a larger GDP than Britain, and was allowing a far larger percentage to its military. An attempt for Britain to match it move for move would have been the recipe for bankruptcy. Chamberlain's view carried a commensurable weight because Britain did not have an actual Ministry of Defense, so it was up to the Treasury to negotiate directly with the air force, navy, and army, the budgets that were required for rearmament. From Chamberlain's point of view, London's key advantage, the so-called fourth arm of its defense in addition to the three services, was its economy. As Chamberlain told Parliament: "wars are won not only with arms and men, they are won with the reserves of reserves and credit." Since Britain could count on its empire to generate raw materials, foodstuffs, and energy resources that were unavailable to Berlin, Rome, and Tokyo, it would have had the eventual upper hand in any long-drawn war of attrition. Therefore, it did not make sense to initiate a large rearmament program, particularly from early on, at the expense of economic viability, because, doing so, in addition to the further burden of maintaining the reconstituted forces, would have undermined Britain's main strength when most needed. Thus, it would have played into the hands of the revisionist powers. As the Treasury put it, "we think we shall probably not be able to [afford rearmament] without bringing down the general economy of this country and thus presenting Hitler with precisely that kind of peaceful victory which would be the most gratifying to him." Accordingly, the best solution appeared to be keeping out of any conflict, while at the same time proceeding to rearm at a pace calculated in such a way as to increase cumulatively Britain's security while at the same time avoiding bankrupting its economy. In the words of Chamberlain's successor as Head of the Treasury, John Simon: "we are in the position of a runner in a race who wants to reserve his spurt for the right time, but does not know where the finishing tape is. The danger is that we might knock our finance to pieces prematurely." Hence, Britain had to time very carefully the precise moment of accelerating its rearmament. Eventually, having marked in 1937 a maximum of £1,570 million for the purpose of rearmament for the following five years, it decided to spend most of this amount over the next two years so as to catch up to Berlin. As a result, by the start of the war, Britain was the power in the better economic position, being able to produce both more aircraft and more tanks than Germany.[27]

Accordingly, Chamberlain did not rule out rearmament—in fact the rearmament program from 1935 to 1940 was the largest that Britain had ever put in place—but instead, prioritized it very stringently. The lion's share of the available budget went to the air force, because Britain remained vulnerable while rearming and needed an immediate deterrent against a surprise attack. After that, the largest piece went to the navy, because it ensured both uninterrupted trade, hence access to imperial resources, and the security of the empire. In the last place was the army, the so-called "Cinderella service," which could not provide immediately for British security, because its chief mission would have been to fight on the continent. As such, if a service had to be sacrificed it was the army: its budget was not only not increased, but further slashed down by half. As the Chairman of the Chiefs of Staff put it in 1937,

> can we, faced with the maintenance of a Navy and Air Force essential for our security as a Nation and as Empire, afford to maintain a Field Force for service overseas? It is submitted that we cannot honestly maintain that such a force is vital to our security.

For Chamberlain, Britain's limited resources "would be more profitably employed in the air, and on the sea, than in the building of great armies." He himself originally envisaged a minimalist British army of around 17 divisions, of which a meager five could have been dispatched to the continent, in stark contrast to the 70 divisions that Britain had deployed in World War I. As things turned out, Chamberlain's Britain made once again a commitment of ground forces to the continent in 1939 by reinstating conscription—but this was a last moment decision once it had become abundantly clear that war was inevitable. A premature reconstitution of the army would have taken away crucial resources from the air force and navy.[28]

Consequently, Chamberlain's measures had a manifest downside: Britain could not have intervened on the European continent to oppose Hitler, even if London had been so inclined, simply because it had no means to do so.[29] For instance, in the Rhineland crisis of 1936, Britain could have only sent to Europe a symbolic force of five brigades, around 25,000 men, to show the flag. As the Chiefs of Staff argued, "if war were to break out, we should not ... be able to mobilize a force with which to reinforce France or Belgium on land for a considerable time." Meanwhile, in the Munich crisis, discussed below, they concluded that "war against Germany, Italy, and Japan simultaneously in 1938 is a commitment which neither the present nor the projected strength of our defense forces is designed to meet." However, Chamberlain's tight-purse stance also came with a major compensating upside: by the time World War II started, Britain had made up for deficiencies where it most counted, in its air force. This would have been impossible without the very careful piece-meal husbanding of resources not just between services, but between various weapon technologies. If the air force had its way, it would have spent most of its vastly increased budget from 1937 on bombers, but this was deemed by the Treasury too expensive—for the price of a bomber, Britain could have built four fighters. Therefore, the money was assigned instead to fighter production, which in conjunction with the invention of radar and air defenses, ensured Britain's victory in the battle of Britain (July–October 1940) against Germany. It should be emphasized that rearmament was pursued in conjunction with passive appeasement. From 1935 to 1937, knowing it was not ready to fight, Britain avoided resorting to a military response to the various aggressive moves of Italy, Japan, and especially Germany. This was affixing a seal of approval in practice to Hitler's rejection of limitations for Germany's military forces, his remilitarization of the Rhineland, a strategic region separating Germany from France, and the unification of Germany and Austria or Anschluss (joining), while criticizing them in rhetoric. All these actions had been prohibited by the peace treaty with Germany signed in the aftermath of World War I. Yet Britain needed more time to rearm and, anticipating that it would not be ready to stand up to Berlin before 1940 at the earliest, had therefore to prove accommodating. To this extent, Chamberlain's grand strategy and his administrative skill proved essential in ensuring Britain's survival.[30]

Why then the public opprobrium? For Chamberlain, playing for time in order to complete rearmament had always been the B-plan, a cautious fallback option in case his plan A did not work out. However, his preference was by far for "removing the danger spots one by one," which meant settling differences with Germany once and for all, thus getting rid of the rationales for hostility and for a wasteful arms race. As he remarked, "if only we could sit down at a table with the Germans and run through their complaints and claims with a pencil, this would greatly relieve all tension." What would such an Anglo-German accord have looked like? Chamberlain believed that Germany's goal was to bring all German populations in neighboring states under the Reich's flag, and, beyond that, the creation of a sphere of influence in Eastern Europe, and, possibly, the retrocession of Germany's old colonies in Africa. These unilateral concessions Chamberlain was willing to make in return for peace, provided that Germany did not force the issue with France's eastern allies, producing therefore a war that nobody wanted and constraining Britain to fight. As Chamberlain reasoned, "give us satisfactory assurances that you won't use force to deal with the Austrians and Czechoslovakians, and we will give you similar assurances that we won't use force to prevent the changes you want, if you can get them by peaceful means." In other words, Chamberlain did not content himself with passive appeasement, but also tried out active appeasement, which then backfired on him.[31]

After the Anschluss, Hitler turned his sights next toward the cession of the Sudetenland, a German-majority inhabited region of Czechoslovakia, which also happened to include the Czech fortifications protecting the republic from an attack by Germany.[32] Essentially, if the Sudetenland was allowed to fall, the rest of Czechoslovakia would have been at the mercy of Germany. To further complicate matters, Czechoslovakia had defense alliances with both France and the Soviet Union. Therefore, the danger was present so that, if Germany attacked Czechoslovakia, France had to step in to defend it, and, then, Britain, despite being not yet ready, had to come in as well, to support France. The British Chiefs of Staff were however highly pessimistic that even an alliance of France and Britain could have at that point defeated Germany. The Soviet Union could not have helped the allies or Czechoslovakia either, as it was separated geographically from its ally by the territories of Poland and Romania. This meant that the Soviet troops could not have reached Czech territory, because Poland and Romania refused them right of passage.[33] Consequently, from Chamberlain's point of view, if the crisis was allowed to escalate into war, this would have been the recipe for a new world war, with Britain and France on the losing side. As Chamberlain concluded, "we could not help Czechoslovakia—she would simply be a pretext for going to war with Germany. That we could not think of, unless we had a reasonable prospect of being able to beat her to her knees in a reasonable time, and of that I see no sign." It is under these dramatic circumstances that Chamberlain moved to conciliate Hitler.[34] Specifically, he took the unprecedented step of resorting to personal diplomacy to diffuse the Sudetenland crisis by visiting Hitler on three occasions in September–October 1938 at his mountain residence at Berchtesgaden, at Godesberg, and, finally, at Munich, in a notorious conference in which the leaders of France and Italy also took part. Basically, the net result of his efforts was that, under threat of invasion by Germany and under warning that neither Britain nor France would lift a finger to help it out should it refuse, Czechoslovakia had to give up the Sudetenland peacefully in return for a guarantee of its remaining territory. Britain's position has forever since been criticized as having thrown Czechoslovakia to the wolves, by reneging on the promise to stand up for it, and having encouraged Germany to ask for even more. To a considerable degree, this is unwarranted: Chamberlain's refusal to fight Germany in 1938 was inevitable so as to gain more time for rearmament.[35]

Nevertheless, Chamberlain could be blamed for two different mistakes having to do with the formulation and the execution of appeasement. Unlike Britain at the turn of the century, which had abandoned regional domination in the Western hemisphere and the Far East, Chamberlain's Britain rejected any trade-offs to change the mind of at least some of its opponents and to gain new allies. In other words, it attempted to conserve the British empire wholesale, without having any priority, or making any concessions to the other great powers. As Barnett writes, this unlimited commitment was not "derived from the logic of political and strategic convenience, but from kinship, from common history and culture, and from loyalty to a common Crown." In other words, sentiment was trumping interest. At the time, Britain could only count on the steady support of France, with the US in isolation. It was therefore imperative for Britain to reduce the number of opponents it was facing, by denying Germany allies, while securing them for itself, in precisely the sort of grand coalition that Churchill was advising. Furthermore, Britain's faraway dependencies could not be adequately defended while London was tied down in Europe. Moreover, they contributed far less to the common defense than their demographic and economic capabilities would have allowed, and, worst of all, drained away much-needed resources from Europe. To exemplify, Britain had made a sizable investment in the 1930s in a fortified base in Singapore and on a navy large enough to intervene to protect it, only to have then Singapore fall to an unforeseen land attack by Japan in 1942. Yet, Britain proved unwilling to reach any understanding involving concessions to either Japan or Italy. In 1922, Britain sought to preserve good relations with the US, Japan's rival in China, by allowing its former alliance treaty to lapse. It was a decision Britain lived to regret further down the road, when Japan emerged not only as a threat to its colonies and to Australia, but also as an ally of Berlin. Meanwhile, in 1935, Britain, France, and Italy, which had an interest in the independence of Austria, constituted the Stresa Front to oppose Hitler. But the very same year, Mussolini decided to invade Ethiopia, a country that Italy had attempted and failed to colonize back in the nineteenth century. Prudence dictated that Britain and France refrain from sanctioning Italy to keep it in their camp—so much was agreed upon in a secret agreement between their foreign ministers, the Hoare-Laval pact. Yet, once the pact was leaked to the press, the government had no recourse in the face of public uproar motivated by Italy's conquest attempt of a member of the League of Nations but to disown the pact and sanction Mussolini, which resulted in Italy also joining forces with Hitler.

But an even worse decision was the refusal of Chamberlain, in the aftermath of the collapse of Czechoslovakia, to consider an alliance with the Soviet Union, despite the manifest interest of Moscow, as seen in the case below, in reaching such an outcome. Unlike Churchill, who in 1941 following Germany's invasion of the Soviet Union, immediately wrote to Stalin to propose an alliance, observing at the same time that if Hitler invaded Hell, he would make at least a favorable reference to the devil in the House of Commons, Chamberlain had previously balked from concluding such a deal with the devil. The price Stalin demanded—a fully fledged political-military alliance between London, Paris, and Moscow and a sphere of influence over the Baltics—was judged to be too steep for two reasons. First, from the perspective of the Western governments, Hitler was not averse to market-run economies, and was representing Germany, with which London and Paris shared a common cultural heritage, and which, therefore, seemed closer in thought and feeling than the Soviet Union. Even after his true colors had shown post-Munich, there was still a lingering belief in Britain that an arrangement over Eastern Europe might be worked out with Hitler, that would prevent a far worse deal with Stalin, and maybe even deflect German aggression away from the west and toward the Soviet Union. Hence, Britain was willing to go through the motions of negotiating with Moscow in 1939, but chiefly in order to pressure Germany to come to terms. A second, and equally strong, reason was that in the 1920s, Britain and France had sought to replace its old alliance with Russia by alliances concluded with the newly independent or substantially enlarged Central and Eastern European states: Poland, Czechoslovakia, Romania, and Yugoslavia. This created a sizable difficulty. Poland and Romania had each gained territory from the Soviet Union, and they feared it far more than they did Germany. Therefore, they were not about to approve of an alliance system that included Stalin. This left Britain and France with the options of either cutting loose its existing allies, which were the very same states that they were defending against Germany, in order to accommodate the Soviet Union, or of keeping the Soviets at arm's length. Consequently, Britain and France did not bid for Moscow's allegiance, forcing Stalin to conclude instead an understanding with Germany at the expense of Britain.

This flaw in the grand strategy's formulation was also compounded by poor execution. Chamberlain left Munich genuinely convinced that he had managed to obtain exactly the grand bargain settlement with Germany that he had wanted in the first place, or, in his words, that he had delivered back "peace for our time." A relieved Britain fell into complacency instead of using the respite to accelerate rearmament even further in the expectation of inevitable war. Instead, Chamberlain stated that "a good deal of false emphasis has been placed ... on rearmament, as though one result of the Munich Agreement has been that it will be necessary to add to our rearmament programs." Chamberlain even angrily rebuffed the demand of his Secretary of State's for the air force for a step-up in rearmament by telling him that "I have made peace!" It should also be said that following his trip to Munich, Chamberlain had been given a hero's welcome, as few other politicians, whether before or since, had ever received—he enjoyed a standing ovation in Parliament and the rarest honor of saluting the crowds from the Buckingham Palace's balcony together with the king. However, Chamberlain was wrong. He had banked on Hitler's ultimately proving a rational decision-maker and choosing the course of action that worked to his best advantage. As Chamberlain later put it, had he dealt with any other decision-maker, it might have well worked. From Chamberlain's perspective, the deal he was offering would have restored in one stroke Germany's power and status, and surely any German leader would have been content with achieving such success without having to resort to armed force. In so doing, Chamberlain ignored his own misgivings that he was actually dealing with a dangerous lunatic. As he confessed, "in spite of the hardness and ruthlessness I thought I saw on his face, I got the impression that here was a man who could be relied upon when he had given his word." Hitler's goals, nevertheless, were far larger, aiming at the complete dominance of the continent and at the annexation of his neighbors in Eastern Europe, not merely exerting influence over them. Conciliating him was an exercise in futility. In March 1939, with the ink not yet dry on the Munich agreement, Hitler proceeded to occupy what was left of Czechoslovakia. In September 1939, when Germany moved against Poland, despite an official British-Polish alliance, Britain, still led by Chamberlain, declared war.[36]

Lying in wait: the Soviet Union and decline

The former Russian Empire was in 1918 at its nadir, having conceded defeat to Germany in World War I, having suffered two consecutive changes of regime, and being engulfed in civil war. Through the peace accord of Brest-Litovsk (1918), it had lost a territory equal in size to double the extent of Germany, an expanse of 1.3 million square miles. This comprised around 50 million people, a quarter of Russia's population, as well as a third of its industrial and agricultural base. Most of the European conquests of Peter the Great and Catherine the Great were lost: Latvia, Estonia, and Lithuania, Russia's section of Poland, Bessarabia, and Finland. While the state, reorganized as the Soviet Union, managed to hold on to Ukraine, this took another war against Poland that extended to 1921. The Soviet Union emerged considerably scathed from this ordeal: the country lost 2 million people in World War I, 2 million in the civil war, 200,000 in the war with Poland, and 15 million in the 1918 global influenza pandemic, overall more than 10% of its population.

As a result, the once formidable Russian bear had turned into a feeble, wounded beast. What had started up as political decline or decay was now spilling into structural decline. Russia had never enjoyed a strong industrial base, and the situation worsened with the Bolsheviks in charge: by 1928, the Soviet Union was producing less than the tsarist government in 1913, while all the great powers produced in excess of their respective World War I manufacturing levels. Even in agriculture, the Soviet Union was falling behind; its production in 1927 was less than a quarter of that on the eve of World War I, going from 9 million to 2 million tons, most of it still conducted by hand, as in medieval times. Meanwhile, the armed forces shrank to fewer than half a million poorly equipped and trained troops. Military spending dropped to 41% of what it had been back in 1913. The Soviet Union only fielded 90 antiquated tanks captured in the civil war; it still relied on cavalry; its artillery had made no progress since 1917; and a third of its soldiers did not even have uniforms. By the 1920s, Russia, which once had competed on equal terms with the strongest powers of the day, feared depredations by non-great powers neighbors such as Poland and Romania.[37] To complete this picture of wretchedness, only Germany officially recognized the Soviet government, while the other great powers treated it as an international pariah.

However, a very different Soviet Union was in place a mere 18 years after. Not only had it recovered the bulk of the territory it had lost, but it also had exceeded the wildest ambitions of the tsars, controlling a sphere of influence extending over the entire eastern half of Europe, the northern Balkans, and well into Germany. The Soviet Union fielded the largest army in the world with around 20,000 tanks, and, although crippled by the war, it was by 1945 an industrial powerhouse, whose sole realistic competitor was the US. For the next half a century, the Soviet Union was to be one of only two superpowers in the world. How was this comeback, unique in history in terms of magnitude and speed, possible? It can be pointed out correctly, that even in its prostrate condition, the Soviet Union still had commensurable latent resources: a huge territory, extensive raw materials, and a large population. But these advantages would have remained useless had it not been for the Soviet grand strategy, whose first principles were articulated by Lenin and who were then elaborated upon and translated into practice with steely ruthlessness by his successor Stalin.

Communism, understood as the synthesis of the thought of Marx and Engels and of Lenin, proceeded from the assumption of an irreconcilable struggle for supremacy between political classes, and, by extension, between countries in which different classes exercised power. For Lenin, coexistence with capitalist states was impossible for two reasons. First, they would have never tolerated the existence of a state that constituted a viable alternative to their socio-economic system: doing so would have inevitably weakened their own legitimacy, and their grip over internal political power. Second, communism could not succeed while living alongside capitalism waiting fruitlessly for revolution to happen. Instead it had to take active steps to overthrow its rival, and, moreover, had to do so relentlessly, using if necessary, dissimulation and violence. Thus, Lenin did not rule out striking temporary tactical compromises with individual capitalist governments whenever doing so advanced Soviet security or economic interests. But deep down, no matter how warm their rhetoric or how accommodating their diplomacy, capitalist states remained the ineluctable enemy, never to be trusted, always animated by a vested interest in undermining the Soviet Union. Conversely, despite officially professing friendship, the Soviet Union would never forget that it was on a historical mission to obliterate one day all capitalist regimes, its current partners included. As Stalin put it, "we cannot have friends; there are immediate enemies and potential enemies." Even official treaty allies represented "the enemies of our enemies, nothing more."[38]

In 1918 and 1919, a host of interventions by Britain, France, the US, and Japan occurred in different parts of the Soviet Union, not to mention the German occupation of part of Ukraine. Had the great powers pressed on, it is unlikely communism would have survived, yet there was no decisive or concerted effort to produce this outcome. Therefore Lenin concluded that

> weak, ruined and crushed, Russia, a most backward country, fought against all the nations, against a league of the rich and powerful states that dominate the world, and emerged victorious. We could not put up a force that was anything like the equal of theirs, and yet we proved the victors. Why was that? Because there was not a jot of unity among them, because each power worked against the other.

Here is where Lenin thought capitalism's fatal flaw lied. No matter how much it hated communism, it cared about profit even more. Greed overrode fear. The more developed the capitalist system, the more the great powers would have embraced economic expansion in search of markets, capital outlets, and raw materials, to the effect that sooner or later these rival imperialist tendencies would have inevitably clashed. This is why Lenin viewed World War I as a gigantic battle between capitalists to increase their share of profit by expropriating each other. And if this was the case, dissension in the capitalist bloc was essential for the success of the Soviet Union, for it meant that if some of the capitalist states saw economic opportunity in cooperating with Moscow, they would have balked at ganging up to destroy it. To quote Stalin, "struggles, conflicts, and wars between our enemies are ... our greatest ally." Therefore, for as long as the Soviet Union managed to strike bargains with at least some of the great powers, it could obtain the economic, technological, military, and diplomatic support it needed to survive. Unwittingly, in so doing, the capitalists were planting the seed of their own destruction by sparing and even strengthening their sworn enemy. As Lenin was reported to have remarked, capitalists would have sold the communists the rope that would be later used to hang them.[39]

Stalin was the one to translate this view of an antagonistic yet divided capitalist world into a functional grand strategy. Stalin's 1931 analysis of the situation of the Soviet Union started from the premise of the state's manifest decline vis-à-vis other great powers: "those who fall behind get beat up. But we do not want to get beat up!" Russia's history, he elaborated, marked a long sequence of defeats and humiliations at the hands of the Mongol Khans, the Ottoman beys, the Swedish barons, the Polish pans, the Anglo-French capitalists, and the Japanese lords, all on account of its relative disadvantage. "One feature of the history of Old Russia was the continual beatings she suffered because of her backwardness. All beat her," concluded Stalin, "because of her backwardness, her military backwardness, cultural backwardness, political backwardness, industrial backwardness, agricultural backwardness. ... We are a hundred years behind the advanced countries." So what was the remedy? As Stalin argued, "we must make good this gap in ten years. Either we do it, or they will crush us." If the Soviet Union wanted to turn things around, it had therefore to reform structurally. Then from a position of strength it could force the other powers to recognize it as an equal or superior great power. If they still refused at that point, it would no longer have mattered:

the Soviet Union could stand up to them. As Stalin advised a delegation from the Mongolian Communist party: "after you strengthen your government and army and raise the economic and cultural level of your people, the imperialist powers will acknowledge you. If they do not, now being strong, you can spit in their faces."[40]

How does a poor and isolated country industrialize and arm at the same time, without concurrently obtaining the necessary capital from abroad? Unlike China in the 1980s, Stalin's Soviet Union had no intention to tamper with the principles of Marxism-Leninism in order to attract foreign investment by encouraging a market-friendly economy. Instead, Stalin chose a solution that harked back to Peter the Great: relying on harsh taxation. This was carried out by the collectivization of agriculture, forcing the peasants to share their property in either collective farms (kolkhoz) or state-run farms (sovkhoz). The Communists had come to power on the promise of distributing the land to the peasants, even though the perpetuation of private property was very much in contradiction to Marxism-Leninism. However, Lenin's efforts to introduce collectivization in the early 1920s had backfired, and he had agreed to a compromise under the New Economic Plan (NEP), in which peasants would keep their land and would deliver to the state a part of their harvest. But by the end of the decade the NEP was coming apart. Peasants did not deliver enough grain to the government on account of not being paid enough, and their small individual lots were ill-suited for carrying out the type of large-scale agricultural production the country needed, basically requiring large farms. The obvious solution was to encourage private enterprise, contacts with the West, and walk away from Marxism-Leninism. But this was anathema to Stalin both on ideological grounds, and, more importantly, as a contradiction of his goal to transform the Soviet Union into a self-contained independent state. Here the resemblance with Peter's Russia ended, because, for all his reliance on squeezing peasants, Peter had never meant to build up a closed country, but instead wanted to bring Russia up to the highest standards of the West, and in so doing make it effectively part of the West. This was not the case for Stalin, who mistrusted and feared Western capitalism. He was not about to make the Soviet Union depend on the good will of its worst enemy. As the Communist Party proclaimed, the goal was "decisively to free industry and the national economy from dependence on foreign countries."[41]

So, instead, Stalin's plan was to force peasants into collectivized farming, de facto creating state-controlled farms by expropriating individual land-owners, large and small. But this measure, while ostensibly presented as a necessary ideological measure, was a means to an end. Once the state was in control of agricultural production, it could set prices as it saw fit, paying peasants a pittance for the food they produced, and selling them manufactured goods at high prices. Furthermore, it would be re-selling the food at four to seven times the rate both abroad and domestically. As Stalin told the Central Committee, the peasants "pay the state not only the usual taxes, direct and indirect, but they also overpay in relatively high prices for industrial goods first of all, and second, they under-receive in prices for agricultural produce." Basically, what Stalin was proposing was to squeeze the population, and, by doing so, raise the capital needed to import technology from the West. He was quite upfront about it too, comparing it to a "tribute" the peasants owed the state: "this is an additional tax on the peasantry in the interests of raising industry, which serves the whole country, including the peasants … something like a supertax, which we are forced to take temporarily … to provide for industry for the whole country."[42]

The consequences of collectivizing by force 80–90% of 100 million people were far-reaching.[43] If this was belt-tightening, it was practiced on a hitherto unprecedented scale in history. Consumption accounted in Western societies for around 80% of national production. During collectivization, the percentage for the Soviet Union dropped to 50%. By 1935, state grain procurement, now easily carried out through the collective farms, accounted for nearly a third of the total budget of the state. Peasants were both promised less labor and mechanization if they joined the kolkhoz; in case promises did not work, they were harassed through fines or petty administrative punishments; and if they still resisted, they would be accused of being enemies of the Soviet power and subjected to beatings or arrest as kulaks. Even so, many preferred to hide their crops, dump them in rivers, or set them on fire, as well as to slaughter the livestock they had to give up to the collective farms. The numbers of farm animals in the Soviet Union dropped dramatically by more than half, and in some cases by three-quarters: horses by 18 million, cattle by 32 million, pigs by 24 million, sheep and goats by 93 million. This caused a severe shortage of not only meat and dairy, but also of animal power, which could not be speedily replaced by new machinery. The biggest land owners, and even the well-to-do peasants, which, for all intents and purposes, was anyone who had employed hired labor or had two or more cows, were accused of being kulaks and marked down for liquidation as a class, even though they represented the country's most industrious and productive farmers. While those who actively resisted collectivization were sent to labor camps, their families and those judged a risk factor were deported to uninhabited areas of Siberia, the Urals, and Kazakhstan. Even those kulaks that were deemed loyal were expropriated, forbidden from joining the kolkhoz, and forced onto the worst land of the outskirts of villages. In total, 7 million people were dekulakized, with 5 million being deported. These combined factors resulted in a sharp reduction in agricultural production, while at the same time grain procurement from collectivized peasants went up more than twice. Simultaneously, the Soviet Union's grain exports went up in the interval from 1927 to 1931 from 0.29 million tons to 5.06 million. The outcome was the spread of an entirely man-made famine through the Ukraine, Kazakhstan, the Volga and Black Earth regions, the North Caucasus, Siberia, and the Urals, with 5 to 7 million dying of starvation, and a further 10 million experiencing the ill-effects of hunger. Many sought to flee the countryside into the cities, only for the government to put into place travel restrictions by introducing an internal passport. The population, which was either forced to subsist on rations in the cities or to starve in the villages, was squeezed even further: the only source of food being the state-run Torgsin shops, which sold food in exchange for gold, precious gems, and foreign currency. Thus, the state managed to acquire even the few valuables the public had managed to hold on to. Vyacheslav Molotov, Stalin's hatchet man and well-reputed for his toughness, reflecting on collectivization, admitted that he had no idea how people withstood it.[44]

Yet, at the same time, the Soviet Union spent 25% of its national production on industrialization. While large swathes of the country were starving to death, the Soviet Union engaged in what was described as an orgy of planning, spending, and construction. This might have been unfeasible if the West had refused to provide the Soviet Union with its technology, but the context was the 1929 Great Depression. Capitalist countries, particularly the US and Germany, could not afford to turn away paying customers. That the Soviet Union had money to burn was not happenstance: the foreign currency it was spending was obtained through Torgsin requisitions and grain and raw materials exports. In fact, since the Depression had lowered the price of these commodities, Moscow was exporting even more. The Soviet Union did not resort to buying specific technologies, but rather bought plants wholesale for reassembly back home, which went directly into operation. Emphasis was placed on heavy industry: iron and steel, metallurgy, tractors and harvesters, automobiles, energy, and armament, and on buying the latest, most advanced technology available.[45] As a result, by the end of the 1928–1933 interval, the first Five Year Plan, the Soviet Union had built more than 1,000 new factories, with some of the largest plants of their kind in the world, based on the Ford assembly line allowing continuous mass production. The Soviet Union's gross industrial production more than doubled, its production of energy and iron ore nearly tripled, and its production of machinery multiplied by seven. Accordingly, by 1935, the Soviet Union had become the most industrialized country in Europe, with an estimated GDP of $334 billion surpassing France with $171, Germany with $275, and Britain with $271, and second in the world only to the US. This was also a very different society from the rural and illiterate tsarist Russia, with an urban population that doubled in size, the creation of an actual working class, and the achievement of a literary rate of nearly 90%. As the country was industrializing, it was at the same time arming up: the size of the armed forces went up nearly three times to 1.4 million troops in 1935. With the newfound industrial power came the ability to develop a large homegrown armament industry, which began to roll out tanks, planes, and artillery at a frenetic pace. By 1935–1936, the Soviet Union had a force of 7,800 tanks of British design, as well as an extensive force of more than 4,000 warplanes. Effectively, the Soviet Union was both outspending and outproducing every single other great power in acquiring weapons (only Germany eventually surpassed it preceding World War II). To sum up: despite and, in fact, largely due to being built on the tears, blood, and bones of its population, Stalin's structural reforms had succeeded in making the Soviet Union the owner of the second economy and, up to the late 1930s, the largest army in the world.[46]

But reform had to be also tempered by caution. Most of the Communist revolutionary leaders of 1917, Lenin included, had been romantics at heart, who believed that the capitalist world was on the brink of revolution. The Soviet Union only needed to hang on and provide would-be revolutionaries with the requisite logistical support through the Third Communist International or Comintern and soon enough it would have been surrounded by friendly like-minded regimes. However, Stalin did not share this faith. By the late 1920s, the Soviet efforts to set up communist governments in Hungary, Germany, Poland, and China had ended in abject defeat and repression. Moreover, as Kennan eloquently wrote, the revolutionary position was self-contradictory: it was as if, on the one hand, the Soviet leaders said to Western countries, "we despise you. We consider that you should be swept from the earth as governments and physically destroyed as individuals. We reserve the right to do what we can to bring this about: to revile you publicly … and to work for your downfall in favor of a communist dictatorship." And yet, on the other hand, they would have added: "but since we are not strong enough to destroy you today … we want you during this interval to trade with us; to finance us; we want you to give us the advantages of full-fledged diplomatic recognition." Of course, this was not going to happen.

Hence, Stalin argued for an alternative point of view: communism's victory depended not on the fortunes of the foreign communist parties, but instead on how the Soviet Union itself fared. It was entirely possible to build communism (socialism) in one country alone, at least at first; if that country were successful, it would have only then used its power to spread the ideology elsewhere. This logic demanded that the interests of foreign parties should have always been secondary to those of the Soviet Union. Accordingly, if the Soviet Union deemed it necessary to cultivate good relations with a particular capitalist or nationalist government, the local Communist Party was supposed to abandon revolutionary agitation and endorse the deal, even if in so doing it assented to its own subservience and even suppression. Even if the local Communists happened to have won power, they still had to place a lower value on the security and prosperity of their own state than on those of the Soviet Union, as the champion of communism and, therefore, the only state that really mattered. While Communists of long standing, and foremost his chief rival Leon Trotsky, accused him of being the gravedigger of the revolution in advocating this line, Stalin was subordinating the Comintern to Soviet national interest.[47]

This meant that, instead of being the engine and command center of the world revolution, the Soviet Union was now playing the part of the brake preventing revolution from taking place. For instance, Communists in China were constrained to cooperate with the Kuomintang in the late 1920s, with the bulk of Soviet aid going to the nationalists, and to accept their own leaders being expelled from government. Even after Chiang Kai-shek cracked down on the Communists with thousands being killed, the Soviet Union had no reluctance to advise in the 1930s that the Communists make up with Chiang and join forces in combating Japan, simply because Chiang was considered indispensable in making China into an independent counterweight to Japan, thus keeping the Soviet Union from having to intervene itself in the Far East. Meanwhile, in the Spanish Civil War that opposed the nationalist forces under General Franco, supported by Fascist Italy and Nazi Germany, and a republican coalition of various forces on the left, the Soviet Union delayed sending out military support, so as not to alarm the West with the specter of a Communist offensive. When Stalin ultimately sent military equipment to Spain, he did so chiefly to increase pressure on Britain and France for not seeking Soviet support in the face of the aggressive international actions of Italy and Germany throughout 1936. Even so, Stalin kept military aid limited (fewer than 1,500 troops compared to 19,000 Germans and 80,000 Italians). As it turned out, Stalin became so suspicious that Spain might come under the sway of a Trotskyist, Anarchist, or Socialist left that it refused to commit further forces and war material to the war, preferring to hand over victory to Franco. But the key caution signal was the evisceration of the Comintern. In effect, capitalist states would always be suspicious of a Soviet government that not only called openly for their overthrow, but also sponsored an agency designed to produce this result. A toothless, ineffective Comintern, fully under Stalin's control, was the best indication that communism had "mellowed," and that it was possible, therefore, to conduct trade, investment, and diplomatic relations with the Soviet Union. As a result, the Comintern was transformed into a hollow organization: after Stalin assumed power, it only met in congress twice, in 1928 and 1935. In 1943, Stalin dissolved it. Thus, socialism in one country allowed the Soviet Union to become stronger without rubbing the great powers the wrong way.[48]

Stalin did not stop here. What would the Soviet Union do once it had regained its strength? He did not intend a Deng Xiaoping-like policy of continually appearing non-threatening to the West, but neither did he seek to call immediately for global revolution. Instead, his recommendation was to wait for the proper opportunity to present itself. As Lenin predicted, the imperialists would inevitably quarrel and go to war with each other, the perfect moment for the Soviet Union to pick the winner, by joining with whatever side offered it more advantages. As Stalin put it as early as 1925: "our banner, as of old, remains the banner of peace, but if war begins, then we must not sit with folded arms—we must act, but act last. And we will act in order to throw the decisive weight on the scales, a weight that must be dominant." As the capitalists fought each other, the Soviet Union was going to sit apart from the fray, helping various sides in turn in exchange for advantages, all the time gaining in strength while its opponents were busy weakening each other. In effect, after concluding the non-aggression pact with Germany, Stalin reflected that

> a war is on between two groups of capitalist countries ... for the redivision of the world, for the domination of the world! We see nothing wrong in their having a good fight and weakening each other. ... We can maneuver, pit one side against the other to set them fighting with each other as fiercely as possible. The non-aggression pact is to a certain degree helping Germany. Next time we'll urge on the other side.[49]

To a certain extent, the Soviet Union's aloofness was not the result of design, but a consequence of the distrust and neglect of Western powers. In the 1920s, with the Soviet Union substantially weakened and committed to bringing down their regimes, it made little sense for great powers to seek alliances with Moscow.[50] As a result, in the 1930s, alignments had already been drawn, with the Soviet Union left out. Specifically, with the coming to power of Hitler in 1933, there was clear antagonism between Germany, as the chief revisionist power, and France and Great Britain as the defenders of the status quo in Europe. But by that time the Soviet Union had become once again a formidable power, whose support could have decided the upcoming conflict. The question to be answered was which one of the two sides was going to present the better offer for an alliance to Stalin. As far as the Soviet Union was concerned, there was little to prefer between the status quo powers or the Nazis. On the one hand, the latter were virulently anti-Communist and, in his book, *Mein Kampf*, which Stalin had read cover to cover, Hitler had spoken at length about the *Drang nach Osten* (drive toward the east) project to expand against the Slavs. But, on the other hand, Great Britain was the chief capitalist power, and, as such, remained by default the supreme enemy of communism. Throughout the 1930s, the Soviet Union's leaders held as an article of faith that Great Britain was enemy number one, who would have liked nothing better than have the Soviet Union pull its chestnuts out of the fire by fighting Germany, while it would have been the one to stand and watch, an assessment by no means farfetched.[51]

British reluctance to conclude an alliance, however, was not matched in Berlin. Germany's chief interest was expansion in Central and Eastern Europe, a goal that put it at odds with the Western powers, which were the patrons of states in the area, but not with the Soviet Union. Even though Hitler's whole revisionist enterprise had been presented as opposing communism, it was actually easier for him to conclude a compromise with Moscow rather than with London and Paris. Hence, the negotiations pursued by Stalin with France and Britain in 1939 pressured Germany in making a bid of its own for Soviet support, or risk being the one left isolated.[52]

For its part, the Soviet Union watched this maneuvering without outwardly flinching ("a riddle wrapped in a mystery inside an enigma," was how Churchill described the Soviet position). A great power running an auction for its allegiance has to maintain dialogue with all sides, which is precisely what the Soviet Union did. While strongly criticizing Germany, Italy, and Japan's transgressions against the international order, and engaging in a formal, but hollow alliance with France, Moscow was at the same time signaling to Berlin that its criticisms and defensive measures were by no means irreversible, and that the Soviet Union and Germany could again work together as they had done in the 1920s. Nor did the Soviet Union supplicate either side for substantial alliance: it let its suitors come to it, rather than chase after them, in the knowledge that any signal of impatience or nervousness would weaken its bargaining position. Other sides would have been alerted that the Soviet Union was so desperate for an alliance that it might have agreed to any terms. Yet maintaining this façade of calm and confidence from 1935 to 1939 was not easy—Stalin was perfectly aware that the great powers could patch up their differences behind his back, and deflect Germany east instead of west, a case in which made Moscow contemplate a possible war on two fronts, against Germany and Japan.[53] In other words, the Soviet Union chiefly feared that it might itself fall victim to the kind of tactics that it was itself using against the West. As Stalin put it:

> the policy of non-intervention [by the Western powers] means conniving at aggression. … The policy of non-intervention reveals an eagerness, a desire not to hinder the aggressors in their nefarious work … not to hinder Germany, say, from… embroiling herself in a war with the Soviet Union; to encourage them surreptitiously in this; to allow them to weaken and exhaust one another; and, then, when they have become weak enough to appear on the scene with fresh strength … and to dictate conditions to the enfeebled belligerents.

Of course, there was nothing Stalin could have done to prevent this outcome, other than waiting and gambling on the contradictions between capitalist powers to be able to do onto them as they would have done onto him. Kotkin compares the resulting situation to a street hustler running a three card monte, which involves picking up the one marked card out of three in connivance with an accomplice (the shill) who is allowed to win, and a victim (the mark) that is saddled with the loss. But who was to be the hustler, the shill, and the mark?[54]

As it happened, Stalin emerged the winning hustler. About to take on a Poland that still refused to yield him the city of Danzig and cast away its neutrality, Hitler needed to secure himself from potential Soviet intervention, so as to be able to turn the bulk of his forces against the Western powers that had guaranteed Polish security. Having made a firm commitment to attack Poland, he had no avenue left for retreat. Hitler's decision thus left Stalin in the advantage—since the Soviet Union had still not committed to either side, it was the party that had the most room left to maneuver, because Britain and France had to move to block a Germany now openly aggressive. Consequently, Hitler's back was against the wall: he had to present Stalin with exactly what he wanted, the recovery of the territory that the Soviet Union had lost at Brest Litovsk. Thus, Germany approached the Soviet Union with an offer that the allies could have never matched: "there are no essential contradictions between our countries from the Black Sea to the Baltic. On all problems it is possible to reach agreement … in more concrete terms." Specifically, this démarche materialized in the Ribbentrop-Molotov Pact of August 1939, involving the transfer of the Baltic countries, Bessarabia, and the portions of Poland inhabited by Ukrainians and Byelorussians to the Soviet sphere of influence. So eager was Hitler for an alliance that he was prepared to go even beyond these limits, and promise even further concessions on Turkey, including the old Tsarist objective of the Straits. The Soviet Union regained nearly all that Russia had lost in World War I with minimal expenditure to itself: while German casualties in Poland, both dead and wounded were around 46,000, the Soviet battle deaths were around 700. Moreover, the Soviet Union had been invited, if not petitioned, or even begged to reassert itself: it had not itself overthrown the status quo of 1918, but took advantage of the opportunity created by Germany. In 1940, as Hitler was busy fighting in the west, Stalin moved to not just put in place a Soviet sphere of influence in the areas mentioned by the Ribbentrop-Molotov Pact, but to annex them. These annexations were then rounded up by a short winter war against Finland, which seized areas protecting the access to Leningrad (the former St. Petersburg). A further benefit of the accord was that Japan, Germany's ally, signed a treaty of neutrality with the Soviet Union, so as to free its hands for action against the colonies of the Western Powers in Southeast Asia, laying to rest Soviet concerns over a clash in the Far East. For all intents and purposes, the Russian empire had made a full comeback under the red hammer-and-sickle flag.[55]

Stalin's calculations had worked well up to that point: but he had counted, as most military experts had forecast, on a war of attrition in the west that would have kept Hitler busy. To Stalin's consternation, France was defeated in a matter of months. As Khrushchev recalls, he had rarely seen Stalin in such a high dudgeon that when he learned of the French defeat: "he literally ran around the room and cursed like a longshoreman. He cursed the French, and he cursed the British, asking how they could have let Hitler smash them like that."[56] A victorious Germany could not have been expected to live up to a bargain it had concluded while still being uncertain of the war's outcome. Yet Stalin soon calmed down. After all, Britain was still fighting, and this did not mean only Britain itself but also its huge colonial empire. Moreover, the US had begun providing Britain with economic and military aid, which opened the door to possible American military involvement. So it would have been logical to assume that faced with such odds, Germany would not seek a simultaneous conflict with the Soviet Union. No rational decision-maker in Hitler's position, and certainly not Stalin, would have done otherwise. Accordingly, Stalin counted on striking yet another great bargain with Germany that would advance Soviet interests even further in Hungary, Romania, and the Balkans, in return for the Soviet Union fully joining the Axis. In order to keep Hitler on his toes, he again tried running an auction by engaging in perfunctory negotiations with Britain. But this was, however, completely misreading Hitler. Like Chamberlain had learned to his expense, the Fuhrer was not a normal, rational statesman—he wanted the Soviet Union to accept a position of junior partner to Germany, if it was allowed to exist at all. What he was prepared to concede to Stalin was territory and influence in Asia, not Europe, which was marked as Germany's exclusive dominium. Even what Stalin had obtained so far seemed excessive to Hitler. He had been upset by the Soviet Union turning its sphere of influence, meant as influence over the foreign policy of the countries involved, into straight occupation and rule by the Communists. So the project took the form of a preventive strike to put down for good the upstart Soviet state. In so doing, Germany would gain significant foodstuffs, raw materials, and an extensive area for German resettlement. Moreover, it would also remove Britain's last hope to win the war by eventually turning the Soviet Union against Germany. If Moscow collapsed, so would London's morale. Finally, the Soviet Union appeared particularly vulnerable. Stalin's purges in the late 1930s had eliminated 90% of the officer corps, which left the Red Army, although formidable on paper, commanded, particularly in terms of field officers, by people with little experience. This weakness had been exposed by the war against Finland, in which the small state, despite being ultimately defeated, exacted 131,000 deaths of the invading Soviet troops. The Soviet Union could have been defeated now, but the more Germany waited, the stronger it would have become.[57] Accordingly, Germany launched a massive offensive against the Soviet Union in June 1941, taking Stalin completely by surprise.

Most military observers expected the Soviet Union to collapse swiftly under the German onslaught, as France and Poland had done. Admittedly, in the first few months of war, the Soviet Union lost again the Baltics and Bessarabia, and, in addition large swathes of Belarus, Ukraine, and Russian territory, including key cities such as Minsk, Kiev, Smolensk, and Odessa, a territory comprising about 40% of the Soviet population, and about half of its industrial production area. By the end of the year, German forces had made 3 million Soviet military prisoners, destroyed nearly half of the Soviet planes and tanks, and inflicted Soviet casualties well exceeding 1 million. How did the Soviet Union not only survive, but also go on to win the war under these circumstances? While it had been the flaws of Stalin's grand strategy that had caused the war, it was also this grand strategy that saved the day. Stalin had been right in 1931: crash industrialization prevented the Soviet Union from going under. Even more important than the Soviet stubborn resistance, the revival of nationalism, notably through a "Holy Russia" speech by Stalin calling the people in defense not of communism but of the country, and the advantages conferred by the weather and the terrain, it was the "battle of engines" that won the war. As Stalin put it, "the war will be won by the side that has an overwhelming preponderance in engine production." This was to say that the side that could replace at a faster rate its destroyed tanks, warplanes, and artillery would ultimately win. As the German forces advanced, Stalin proceeded to relocate armament, metallurgy, and automobile factories from endangered territory to the Urals, Kazakhstan, the Volga region, and Siberia. In total, more than 1,500 plants were moved away, together with their machinery and workers: around a half-a-million railway wagons' load. This required a race against the clock, sometimes in as little as two weeks, to create the conditions in the new locations for these plants to start operating as soon as they were moved. The new factory buildings often were not even made of concrete or brick, but out of wood, some of them did not even have roofs. But the moved plants started operating on schedule, turning up tens of thousands of tanks, guns, and warplanes. Simultaneously, the Soviet Union opened up new mines and power plants to replace the raw materials in the territory lost to Germany and mobilized its population, including the women and the young to operate them. The contribution provided by the Soviet Union's new allies, Britain and the United States, to the Soviet war-machine was certainly not negligible. But the bulk of this aid came in after 1943, and amounted for no more than 10–15% of the total Soviet war production. This means that the Soviet Union won because it managed to outproduce Germany, with around 2,000 new tanks and 3,000 warplanes coming out every month. By the end of the war, the total figures of Soviet war material produced stood at 100,000 tanks, 120,000 planes, and 360,000 guns. It is not a coincidence that the Soviet Union experienced its worst military reverses during the time this military-industrial production chain was being set up, from 1941 to mid-1942, since it was still outgunned by the more advanced German army. But once it did go into production, the Soviets began having the upper hand. In every major battle from Stalingrad onwards, the Soviet Union enjoyed a advantage in military equipment over Germany. Once a second front was created in France in June 1944, the final defeat of Hitler was inevitable, an outcome that Stalin had anticipated as far back as 1941.[58]

Initially, accusing Germany's surprise attack, Soviet diplomacy had shown signs of panic—after Churchill proposed an alliance, Stalin pressed for the opening of a second front and even petitioned for the use of British troops on Soviet territory. But once the German offensive petered out at the gates of Moscow in December 1941, and the US entered the war the same month, as a result of Germany declaring war in the aftermath of Pearl Harbor, the Soviet Union regained its confidence and turned its attention to the matter of territorial arrangements in Europe. In the Atlantic Charter of August 1941, Britain and the US had denounced territorial changes that did not accord with the express wishes of the peoples concerned. This provision nullified not only the annexations by Germany, but also impacted the territorial gains that the Soviet Union had made through the Ribbentrop-Molotov pact. Consequently, the Soviet Union sought and obtained from the allies an exception to the rule: while the Axis had to give back seized land, the Kremlin was allowed to keep all that it had obtained, even though manifestly it was against the wishes of the peoples annexed. This proved particularly awkward in the case of Poland, which was to resume its sovereignty following the eventual defeat of Hitler, but would not be able to regain its territorial integrity. Therefore, Stalin simply suggested that Polish frontiers be moved westward: Poland was going to be compensated for its losses in the east with the transfer of East Prussia in the west. The Western Allies had to swallow eventually their previous words since they could not press Stalin on the point, as he enjoyed considerable leverage by being the only power to actually fight the Axis in Europe from 1941 to 1943. Accordingly, since its contribution was essential to the war effort, the Soviet Union could name its price. It never affirmed, like Britain or the US, that it was not going to seek out any territorial aggrandizement, and ended up rounding its territory with the addition of Königsberg in Eastern Prussia, Carpathian Ruthenia from Czechoslovakia, and Sakhalin and the Kuril Islands from Japan.[59]

This might have seemed a fair price to pay for Soviet support in the war: a proportion of 80% of Germany's casualties after all took place on the Eastern front. But Stalin wanted more. Once his demands concerning Ribbentrop-Molotov were recognized, he started thinking about the strategic situation that would emerge once the war ended. From his point of view, capitalist powers were still not to be trusted: they would treat a defeated Germany (and Japan) with "kids' gloves," allowing them to get back on their feet and once again resume their offensive against the Soviet Union. In these circumstances, the best solution was to create a ring of Soviet bases and allied regimes around each country by seeking out a Soviet sphere of influence in Europe and East Asia. Of course, Stalin did not expect the Western powers to just cede him half of Europe. Fairly soon, it became clear that an explicit sphere of influence agreement was not going to be possible, as seen from Chapter Five. It is true that Churchill proposed to Stalin a sphere of influence agreement in October 1944 in Moscow, expressed in percentages, which the Soviet despot casually ticked off, but even Churchill had qualms about what he had just done. As he put it, "might it not be thought rather cynical if it seemed we had disposed of these issues so fateful to millions of people in such an offhand manner?" He then suggested burning the piece of paper. Nor was this initiative followed up—it would have probably been nixed by the US. Nevertheless, Stalin aimed to achieve in practice what he had been denied on a formal base. If the Soviet Union's advancing armies entered a country, no matter whether this had been an ally of Germany, a victim of its occupation, or a neutral party, then it became automatically part of the Soviet sphere. To ensure its allegiance, the Soviet Union would insist at the inclusion of Communists in government, which soon enough would then take over the entire political and economic leadership. Moreover, following Stalin's principle of revolution in one country, such governments would have been entirely subordinated to Moscow instead of pursuing their own independent national interests. As Stalin observed: "this war is not as in the past: whoever occupies a territory also imposes on it his own social system. Everyone imposes his own system as far as his army can reach. It cannot be otherwise." Of course, once a territory had been occupied by the Red Army, it would have been next to impossible for the US and Britain to force it out. Thus, de facto, they would have respected the existence of a Soviet sphere even though they would never acknowledge it. Stalin, in turn, was not interested in interfering in the Western sphere of influence—the area under British and American military control—as far as he was concerned, he reciprocated the favor he himself was being extended. But until the end of the war, Europe was up for grabs, and the last two years of war saw a frenzied Soviet effort to advance as far as possible in the west, with a multitude of potential expansion targets, including Denmark, Norway, Sweden, and Switzerland (Stalin had also expressed an interest in getting Soviet troops all the way to Paris, as in the times of Tsar Alexander). Hence, at a time where the Western Allies were still busy fighting the war with the Axis and considered how to prolong cooperation with the Soviet Union under the auspices of the nascent UN, the Kremlin had already adopted a Cold War mentality. If the capitalist powers were going to try to double-cross the Soviet Union, they would find out that they had been double-crossed in turn.[60]

Of all the grand strategists covered by this book, there is likely none as controversial as Stalin. His measures left millions dead and were responsible for pain, suffering, misery, and despair on an unimaginable scale. When the cost is the loss through famine, purges, forced resettlement, and war of more than one in ten of your country's citizens, it is fair to ask if the benefits in power and status could not have been achieved with considerably less sacrifice. But this may be asking the right question in a wrong way. It is possible that the Soviet Union could have made a comeback over time by opening itself economically and politically to the West. But this was unacceptable from the point of view of its Communist ideology, and therefore could not have been realistically implemented. Moreover, such a comeback would have taken place only slowly and the Soviet Union, faced as it was with the likelihood of attack from one or several great powers, precisely the situation that occurred in World War II, had to rely on a fast transition through forceful methods from a backward economy, society, and military into being once more a powerhouse. As a Russian saying attributed to Catherine the Great goes, victors are not judged.

Notes

1 Levy and Ripsman, "Preventive War that Never Was"; Treisman, "Rational Appeasement"; Rock, *Appeasement*, 12–15.
2 Gilpin, *War and Change*, 192.
3 Rock, *British Appeasement*, 25. Wars normally begin with a crisis, an event triggered by another power's action that threatens the national interest and reputation, presents the risk of war, and forces a decision under time constraints. If crises are prevented or managed, then war may not break out. A particular type of crisis is the justification of hostility crisis, in which a state bent on war employs the pretext of a real or perceived offense or provocation to enflame public opinion and begin a war it had already decided to launch. A meek foreign policy is designed precisely to make it very difficult for a would-be aggressor to find such a suitable pretext. Lebow, *Between Peace and War*, 24–37.
4 Rock, *British Appeasement*, 25; George and Craig, *Force and Statecraft*, 157, 246–7; Kennedy, "Tradition of Appeasement," 15–16.
5 Gilpin, *War and Change*, 193–4.
6 This may seem similar to semi-detachment, but it is not the same grand strategy. Semi-detachment allows the practitioner to choose its level of involvement in the core, scaling it up and down from intervention to passivity. In other words, semi-detachment is not necessarily opting for being continually detached from the events taking place in the core by conciliating great powers there.
7 Levy and Ripsman, "Preventive War that Never Was," 39–43; Layne, "Security Studies," 408–9.
8 Machiavelli, *Discourses*, Book II: 14, 255; Treisman, "Rational Appeasement," 349–51; Rock, *Appeasement*, 7–8.
9 Machiavelli, *Discourses*, Book II: 14, 255; Snyder and Diesing, *Conflict among Nations*, 186–9; Press, *Calculating Credibility*; Mercer, *Reputation and International Politics*, 39–43; Jervis, *Perception and Misperception*, 90.
10 Ibid., 187–91.
11 Taliaferro, "Neoclassical Realism and Resource Extraction: State Building for Future War," 213–14; Christensen, *Useful Adversaries*, 11–28.
12 Wetzel, *Duel of Giants*; Echard, *Napoleon III*.
13 Christensen and Snyder, "Chained Gangs and Passed Bucks"; Schweller, *Deadly Imbalances*, 73–4. Also see on hedging understood as the effort to maintain good relations simultaneously with rival powers without fully committing to either, thus leaving one's options open see Lim and Cooper, "Reassessing Hedging"; Tessman, "System Structure and State Strategy."
14 CATO, *Guilty Men*; Morgenthau, *Politics Among Nations*, 68–71.
15 Kennedy, *Rise and Fall of Great Powers*, 200–1, 203, 231, 224–32; Maddison, *Monitoring the World Economy*, 106–7, 182–3; Perkins, *Great Rapprochement*, 121–2, 122–6; Friedberg, *Weary Titan*, 67; Hobsbawm, *Age of Empire*, 59.
16 Sanderson, "Memorandum by Lord Sanderson," 430; Joseph Chamberlain, "Opening Speech at Colonial Conference, London, June 30, 1902" in Amery, *Life of Joseph Chamberlain*, 31.

17 Padfield, *Great Naval Race*, 184.
18 As Selborne put it: "if I am asked why I select France and Russia, and none others, as the two powers, which are to regulate our standard, I would submit that a naval war with France [the second power], or with France and Russia is less improbable than any other naval war which we can foresee." "Cabinet Memorandum by Lord Selborne, 16 November 1901" in Boyce, *Crisis of British Power*, 129–36, 134; Friedberg, *Weary Titan*, 172–3.
19 Ibid., 175–8; Boyce, ed., *Crisis of British Power*, 123–6; Monger, *End of Isolation*, 49–50.
20 Friedberg, *Weary Titan*, 204; Bourne, *Britain and the Balance of Power*, 385; Kennedy, *Rise and Fall of British Naval Mastery*, 214; Beloff, *Imperial Sunset*, 87.
21 In relation to the US, see Bourne, *Britain and the Balance of Power*; Perkins, *Great Rapprochement*; in relation to Japan, see Murashima, "Opening of the Twentieth Century and the Anglo-Japanese Alliance," 157–96.
22 Marder, *Anatomy of British Sea Power*, 442, fn. 1, 449–52; Padfield, *Great Naval Race*; Monger, *End of Isolation*; Robbins, *Edward Grey*.
23 Howard, *Continental Commitment*, 75; Barnett, *Collapse of British Power*, 123, 202, 208, chap. 4.
24 Kennedy, *Rise and Fall of Great Powers*, 281–2, 316–9; Howard, *Continental Commitment*, 89–94; Peden, *British Rearmament*, 3–9; Alford, *Depression and Recovery*.
25 Peden, *British Rearmament*, 67–70, 109–12; Layne, "Security Studies," 404–8; Howard, *Continental Commitment*, 120.
26 Adams, *British Politics and Foreign Policy in an Age of Appeasement*, 157–8; Carr, *Britain*.
27 Peden, *British Rearmament*, 14, 65–6, 103–4; Shay, *British Rearmament*, 78–81; Feiling, *Life of Neville Chamberlain*, 254, 252–3, 256, 319. Another consideration in the mind of the British decision-makers particularly from 1932 to 1935 had to do with public resistance to rearmament, which might have pushed it to embrace communism. However, in 1935 the Tories won the last election before the war on a platform that stressed rearmament, although a moderate version of it. Schweller, *Unanswered Threats*, 69–75; also see Levy and Ripsman, "The War that Never Was," 59–62.
28 Peden, *British Rearmament*, chap. 4, esp. 124–5, 127–8, 137–8; Howard, *Continental Commitment*, 104–22.
29 Chamberlain, following Canning, was very much against bluffing: making threats that one did not intend or did not have the power to implement. As he put it, "over and over again, Canning lays it down that you should never menace unless you are in a position to carry out your threats, and … we are certainly not in a position in which our military advisers would feel happy in undertaking to begin hostilities." Fieling, *Life of Neville Chamberlain*, 360.
30 Peden, *British Rearmament*, 127–34; Levy and Ripsman, "Preventive War that Never Was," 55–7; Fieling, *Life of Neville Chamberlain*, 319.
31 Ibid., 333; Adams, *British Politics and Foreign Policy*, 68.
32 At the time, Czechoslovakia's ethnic makeup was of 7 million Czechs, 3 million Germans, 2 million Slovaks, and 1 million of other nationalities. Yet, despite being a bare majority, the Czechs held the key offices in the state.
33 Even if passage could have been arranged, the alliance treaty stipulated that the Soviet Union would intervene to help Czechoslovakia only if France intervened first. As France signed the Munich agreement of September 1938, and then did not respond militarily to the German occupation of Czechoslovakia the following year, there was no justification for any Soviet intervention, although Stalin was prepared for military action. Kotkin, *Stalin: Waiting for Hitler*, 565, 596.
34 Fieling, *Life of Neville Chamberlain*, 348; Adams, *British Politics and Foreign Policy*, 39–53, chaps. 5–6.

35 It is highly doubtful that an alternative strategy of confrontation, as the one Churchill advocated, would have worked any better. In fact, it might have well led to disaster: there are few indications that Hitler would have been deterred and available evidence indicates he would have fought over Czechoslovakia. Gerhard Weinberg, *The Foreign Policy of Hitler's Germany: Starting World War II, 1937–1939* (Chicago: University of Chicago Press, 1980).

36 Adams, *British Politics and Foreign Policy*, 130, 132; Fieling, *Life of Neville Chamberlain*, 367; Layne, "Security Studies and the Use of History," fn. 64, 413.

37 Kotkin, *Stalin: Paradoxes of Power*, 258, 312, 619, 662–3; Kotkin, *Stalin: Waiting for Hitler*, 21–2; Davies, *Socialist Offensive*, 4–18; Davies, *Soviet Economy in Turmoil*, 13–29. In the aftermath of World War I, the Soviet Union also had lost control over Georgia, Azerbaijan, and Armenia, but this was only temporary: by the early 1920s, Moscow had reasserted control.

38 Ulam, *Expansion and Coexistence*, 18–26; Kotkin, *Stalin: Waiting for Hitler*, 413, 21.

39 Ulam, *Expansion and Coexistence*, 70; V. I. Lenin, "Report on the International Situation and the Fundamental Tasks of the Communist International, July 19 1920," at https:/ /www.marxists.org/archive/lenin/works/1920/jul/x03.htm#fw1; Lenin, *Imperialism*; Kotkin, *Stalin: Waiting for Hitler*, 17; Kotkin, *Stalin: Paradoxes of Power*, 557–8. Kennan thought that Stalin followed a divide and rule approach, by letting his opponents, domestic or foreign, fight among themselves and so waste their strength, while he conserved his. Kennan, *Russia and the West*, 239–40.

40 Roberts, *Stalin's Wars*, 22; Kotkin, *Stalin: Waiting for Hitler*, 73–4, 195, 494–5, 572.

41 Nove, *Economic History*, 129–34, 138–44, 180.

42 Kotkin, *Stalin: Paradoxes of Power*, 710–1.

43 There is debate as to whether the famine produced by these measures had been intentionally caused by Stalin, as an instrument of exterminating class and national opponents, or was an unintended by-product, which is the currently favored view by scholarship, because Stalin mistakenly believed that the introduction of tractors and harvesters would result in a larger agricultural production that would compensate for the shortages. What is known however was that Stalin was constantly informed about the famine, but characteristically blamed it on saboteurs and class enemies, who stole the grain necessary to feed the country and the military. He also took measures to distance himself from collectivization abuses, denouncing those activists who, "dizzy with success," resorted to coercing people to join the kolkhoz. Furthermore, he took measures, although belatedly, to reduce seizures of grain and even resorted to imports, as well as bought new livestock from abroad. Conquest, *Harvest of Sorrow*, 322–9; Kotkin, *Stalin: Waiting for Hitler*, 127–9.

44 The most comprehensive discussion is Conquest, *Harvest of Sorrow*. Also see Nove, *Economic History*, 150–8, 163–6, 170, 202–3; Kotkin, *Stalin: Waiting for Hitler*, 127–8, 131; Kotkin, *Stalin: Paradoxes of Power*, 724; Kennedy, *Rise and Fall of Great Powers*, 323–5.

45 This does not mean that the First Five Year Plan was a complete success. Its targets had been highly unrealistic, the statistics reported had been inflated or manipulated to meet targets, and in the drive to rush through production, the machines experienced a higher rate of failure and accidents, as well as having a shorter lifespan than their Western counterparts. Nove, *Economic History*, 178–87; Kotkin, *Stalin: Waiting for Hitler*, 45, 48, 70–3.

46 Ibid., 32, 49–50, 131–2, 91, 188, 290; Nove, *Economic History*, 183, 185, 201–3; Angus Maddison, "Statistics on World Population, GDP and Per Capita GDP, 1–2008 AD," at http:// www.ggdc.net/maddison/; Kennedy, *Rise and Fall of Great Powers*, 323–5; The Correlates of War Project, *National Military Capabilities*, v. 5.0 at http:// www.correlatesofwar.org/data-sets/national-material-capabilities.

47 Kotkin, *Stalin: Paradoxes of Power*, 555, 615, 557; Kennan, *Russia*, 176; Ulam, *Coexistence and Expansion*, 112–25.

48 Kotkin, *Stalin: Paradoxes of Power*, 626–33; Kotkin, *Stalin: Waiting for Hitler*, 248, 318–20, 337–9, 345–6, 350–2, 361–7, 405–6, 409–11, 431, 615–16; Ulam, *Expansion and Coexistence*, 186–7, 228–33, 346–7; Haslam, *Soviet Union*.

49 Kotkin, *Stalin: Paradoxes of Power*, 555, 557, 615; Roberts, *Stalin's Wars*, 36; Kennan, *Russia and the West*, 176.

50 A partial exception was Germany. Because it had been forbidden from owning or developing advanced weaponry by the Versailles Peace Treaty in the aftermath of World War I, Germany sought to bypass the restrictions by building and testing this arsenal in a place where no one would look: in the Soviet territory. In exchange, the Kremlin got both diplomatic recognition and access to the new weapon designs. This arrangement agreed upon at Rapallo in 1922 lasted until the early 1930s, but should not be construed as an alliance. Instead, German leaders prior to Hitler wanted to eventually come to an agreement with France and Great Britain.

51 Kotkin, *Stalin: Waiting for Hitler*, 168, 274, 292, 581–2, 608, 643, 681–2.

52 Haslam, *Soviet Union*, 87–94, 98–9, 158–9, 165–8, 206–7, 213–29; Kotkin, *Stalin: Waiting for Hitler*, 590–3, 612–17, 621–3, 633–6, 647–9, 652–3, 655–7, 673–5.

53 This agonizing waiting was probably at least in part responsible for Stalin's Great Terror—the purge, arrest, torture, and execution of between a third and half of the leaders and staff of the Communist Party; the officer corps; the economic managers; the local administration; the Comintern; and, for good measure, the secret police on accusations of espionage and sabotage, a figure exceeding 1 million people. The most common explanations of the Great Terror have to do with a) a preventive action by Stalin, who eliminated anyone that might have sided with the enemy in case of a war, and b) his concern for his political power—for in case of a war, the subjects of the purge might have acted to unseat him. Kotkin, *Stalin: Waiting for Hitler*, chaps. 7–9; Conquest, *Great Terror*.

54 Kotkin, *Stalin: Waiting for Hitler*, 555–79, 632–3, 659–6; Roberts, *Soviet Union*, 65–8, 72–3, 79–82, 89–91; Haslam, *Soviet Union*, 77, 95–6, 153–4, 205–6. Timothy Crawford similarly argues that both the Soviet Union and Germany were employing wedge tactics in order to separate their potential enemies: the Soviet Union neutralized Japan, and also broke the potential of a British-German understanding; at the same time Germany prevented the constitution of an alliance between Britain, France, and the Soviet Union. Crawford, "Powers of Division," 246–78, 259–63.

55 Kotkin, *Stalin: Waiting for Hitler*, 687–9, 770–5, chap. 12; Roberts, *Soviet Union*, 94–5, 104–21.

56 Kotkin, *Stalin: Waiting for Hitler*, 768–9.

57 Ibid., 782–6, 799–800, 804–9, 817–19; Roberts, *Soviet Union*, 137–8. However, Stalin believed that war with Germany might come further down the road, in 1942 or later.

58 Werth, *Russia at War*, 213–16, 219–24, 620–3, 625–7; Roberts, *Stalin's Wars*, 109–10, 162–4.

59 Roberts, *Stalin's Wars*, 114–15, 167, 185–7; Mastny, *Russia's Road to the Cold War*, 41–3, 50, 129–32; Ulam, *Expansion and Coexistence*, 317–21, 352–4.

60 Mastny, *Russia's Road to the Cold War*, 212–24, 239–53, 267–80; Roberts, *Stalin's Wars*, 180–91, 217–25, 237–45, 246–8, 272–9.

Bibliography

Adams, RJQ. *British Politics and Foreign Policy in an Age of Appeasement, 1935–1939*. Stanford: Stanford University Press, 1993.

Alford, BWE. *Depression and Recovery: British Economic Growth, 1919–1939*. London: Macmillan, 1972.

Amery, Julian. *The Life of Joseph Chamberlain*, vol. 5. London: Macmillan, 1969.

Barnett, Corelli. *The Collapse of British Power*. Phoenix Mill: Sutton Publishing, 1997.

Beloff, Max. *Imperial Sunset*, vol. 1. New York: Knopf, 1970.

Bourne, Kenneth. *Britain and the Balance of Power in North America, 1815–1908*. London: Longmans, 1967.

Boyce, George. *The Crisis of British Power: The Imperial and Naval Papers of the Second Earl of Selborne, 1895–1910*. London: Historian's Press, 1990.

Carr, EH. *Britain: A Study of Foreign Policy from the Versailles Treaty to the Outbreak of the War*. London: Longmans, 1939.

CATO. *Guilty Men*. London: Victor Gollancz, 1940.

Christensen, Thomas. *Useful Adversaries: Grand Strategy, Domestic Mobilization, and Sino-American Conflict, 1947–1958*. Princeton: Princeton University Press, 1996.

Christensen, Thomas and Jack Snyder. "Chained Gangs and Passed Bucks: Predicting Alliance Patterns in Multipolarity." *International Organization* 44 (Spring 1990): 137–68.

Conquest, Robert. *The Great Terror: A Reassessment*. New York: Oxford University Press, 2008.

Conquest, Robert. *The Harvest of Sorrow: Soviet Collectivization and the Terror-Famine*. London: Hutchinson, 1986.

Crawford, Timothy. "Powers of Division: From the Anti-Comintern to the Nazi-Soviet and Japanese-Soviet Pacts, 1936–1941." In *The Challenge of Grand Strategy: Great Powers and the Broken Balance between the World Wars*, edited by Jeffrey Taliaferro, Norrin Ripsman, and Steven Lobell. Cambridge: Cambridge University Press, 2012.

Davies, RW. *The Socialist Offensive: The Collectivization of Soviet Agriculture, 1929–1930*. New York: Palgrave Macmillan, 1989.

Davies, RW. *The Soviet Economy in Turmoil, 1929–1930*. New York: Palgrave Macmillan, 1989.

Echard, William. *Napoleon III and the Concert of Europe*. Baton Rouge: Louisiana State University Press, 1983.

Feiling, Keith. *The Life of Neville Chamberlain*. London: Macmillan & Co., 1946.

Friedberg, Aaron. *The Weary Titan: Britain and the Experience of Relative Decline, 1895–1905*. Princeton: Princeton University Press, 1988.

George, Alexander and Gordon Craig. *Force and Statecraft: Diplomatic Problems of Our Time*. New York: Oxford University Press, 1995.

Gilpin, Robert. *War and Change in World Politics*. New York: Cambridge University Press, 1981.

Haslam, Jonathan. *The Soviet Union and the Struggle for Collective Security in Europe, 1933–1939*. New York: Palgrave Macmillan, 1984.

Hobsbawm, Eric. *Age of Empire, 1875–1914*. London: Weidenfeld & Nicolson, 1987.

Howard, Michael. *The Continental Commitment: The Dilemma of British Defense Policy in the Era of the Two World Wars*. London: T. Smith, 1972.

Jervis, Robert. *Perception and Misperception in International Politics*. Princeton: Princeton University Press, 1975.

Kennan, George. *Russia and the West Under Lenin and Stalin*. New York: Mentor, 1961.

Kennedy, Paul. *The Rise and Fall of British Naval Mastery*. London: Allen Lane, 1976.

Kennedy, Paul. *The Rise and Fall of Great Powers: Economic Change and Military Contest From 1500 to 2000*. New York: Random House, 1987.

Kennedy, Paul. "The Tradition of Appeasement in British Foreign Policy, 1865–1939." In *Strategy and Diplomacy, 1870–1945: Eight Case Studies*, edited by Paul Kennedy. Boston: Allen & Unwyn, 1983.

Kotkin, Stephen. *Stalin: Paradoxes of Power, 1878–1928*. New York: Penguin Books, 2014.

Kotkin, Stephen. *Stalin: Waiting for Hitler, 1929–1941*. New York: Penguin Press, 2017.

Layne, Christopher. "Security Studies and the Use of History: Neville Chamberlain's Grand Strategy Revisited." *Security Studies* 17, no. 3 (2008): 397–437.

Lebow, Richard Ned. *Between Peace and War: The Nature of International Crisis*. Baltimore: Johns Hopkins University Press, 1981.

Lenin, VI. *Imperialism: The Highest Stage of Capitalism*. Moscow: Cooperative Publishing Society of Foreign Workers in the USSR, 1934.

Levy, Jack and Norrin Ripsman. "The Preventive War that Never Was: Britain, France and the Rise of Germany in the 1930s." *Security Studies* 16, no. 1 (2007): 32–67.

Lim, Daniel and Zack Cooper. "Reassessing Hedging: The Logic of Alignments in East Asia." *Security Studies* 24, 4 (2015): 696–727.

Machiavelli, Niccolò. *Discourses on the First Decade of Titus Livius*. New York: Dover Publications, 2007.

Maddison, Angus. *Monitoring the World Economy*. Paris: Development Center of the Organisation for Economic Co-operation and Development, 1995.

Marder, AJ. *Anatomy of British Sea Power: A History of British Policy in the Pre-Dreadnought Era, 1880–1905*. New York: Knopf, 1940.

Mastny, Vojtech. *Russia's Road to the Cold War: Diplomacy, Warfare, and the Politics of Communism, 1941–1945*. New York: Columbia University Press, 1979.

Mercer, Jonathan. *Reputation and International Politics*. Ithaca: Cornell University Press, 1995.

Monger, George. *The End of Isolation: British Foreign Policy, 1900–1907*. London: Thomas Nelson and Sons, 1963.

Morgenthau, Hans. *Politics Among Nations: The Struggle for Power and Peace*. New York: Knopf, 1968.

Murashima, Shigeru. "The Opening of the Twentieth Century and the Anglo-Japanese Alliance, 1895–1923." In *The History of Anglo-Japanese Relations*, vol. 1, edited by Ian Nish and Yoichi Kibata. New York: St. Martin's Press, 2000.

Nove, Alec. *An Economic History of the USSR*, 2nd edn. New York: Penguin Books, 1989.

Padfield, Peter. *The Great Naval Race: The Anglo-German Naval Rivalry, 1900–1914*. New York: David McKay Company, 1974.

Peden, CG. *British Rearmament and the Treasury, 1932–1939*. Edinburgh: Scottish Academy Press, 1939.

Perkins, Bradford. *The Great Rapprochement: England and the United States, 1895–1914*. New York: Athenaeum, 1968.

Press, Darryl. *Calculating Credibility: How Leaders Assess Military Threats*. Ithaca: Cornell University Press, 2005.

Robbins, Keith. *Sir Edward Grey: A Biography of Lord Grey of Fallodon*. London: Cassell, 1971.

Roberts, Geoffrey. *Stalin's Wars: From World War to Cold War, 1939–1953*. New Haven: Yale University Press, 2008.

Rock, Stephen. *Appeasement in International Politics*. Lexington: University Press of Kentucky, 2000.

Rock, William. *British Appeasement in the 1930s*. New York: W. W. Norton, 1977.

Sanderson, Thomas. "Memorandum by Lord Sanderson." In *British Documents on the Origins of the War, 1898–1914*, vol. VI, edited by GP Gooch and Harold Temperley. London: His Majesty's Stationery Office, 1928.

Schweller, Randall. *Deadly Imbalances: Tripolarity and Hitler's Strategy of Conquest*. New York: Columbia University Press, 1998.

Schweller, Randall. *Unanswered Threats: Political Constraints on the Balance of Power*. Princeton: Princeton University Press, 2006.

Shay, Robert Paul. *British Rearmament in the 1930s: Politics and Profits*. Princeton: Princeton University Press, 1977.

Snyder, Glenn and Paul Diesing. *Conflict Among Nations: Bargaining, Decision-Making, and System Structure in International Crises*. Princeton: Princeton University Press, 1977.

Taliaferro, Jeffrey. "Neoclassical Realism and Resource Extraction: State Building for Future War." In *Neoclassical Realism, The State, and Foreign Policy*, edited by Norrin Ripsman, Steven Lobell, and Jeffrey Taliaferro. Cambridge: Cambridge University Press, 2009.

Tessman, Brock. "System Structure and State Strategy: Adding Hedging to the Menu." *Security Studies* 21 (2012): 192–231.

Treisman, Daniel. "Rational Appeasement." *International Organization* 58 (Spring 2004): 345–73.

Ulam, Adam. *Expansion and Coexistence: The History of Soviet Foreign Policy*. New York: Praeger, 1968.

Weinberg, Gerhard. *The Foreign Policy of Hitler's Germany: Starting World War II, 1937–1939*. Chicago: University of Chicago Press, 1980.

Werth, Alexander. *Russia at War, 1941–45*. New York: E. P. Dutton & Co., 1964.

Wetzel, David. *A Duel of Giants: Bismarck, Napoleon III, and the Diplomacy of the War of the Franco-Prussian War*. Madison: University of Wisconsin Press, 2012.

7 Great powers' grand strategies today and tomorrow

What are the present grand strategies of the great powers? Do they have grand strategies in place? What are the advantages and pitfalls of their current strategic course? And which grand strategy should they follow in the future?

This book argues that the assessment of grand strategy should begin by asking what sort of great power one is dealing with. Is the great power in question seeking to increase its power and status, maintain them, or to prevent, minimize, or recover their loss? In other words, is this a rising, status quo, or declining power? Identifying the type of great power is both a matter of observing objective trends in material dimensions, foremost demographics, military spending, and GDP, Tables 7.1, 7.2, and 7.3; and of qualitatively monitoring the perceptions of the great power in question as well as of the other actors in the system as to where it ranks.[1] Then, depending on the answer to the above question, one should consider the possible strategies available for that respective type of power.

Hence, applying this basic formula to the existing seven great powers (the US, China, Russia, Germany, Japan, France, and Great Britain) present in the post-Cold War results in classifying the US, Germany, Japan, France, and Great Britain as status quo powers, China as the lone rising power in the club, and Russia as a declining power making a comeback. In what follows, owing to space constraints, this chapter will examine the grand strategy of the great powers that arguably occupy the first ranks: the US, China, and Russia.[2]

The United States

In the post-Cold War, the US foreign policy establishment was kept busy by the so-called Kennan sweepstakes, an effort to come up with a "bumper sticker" that would constitute a worthy heir to Kennan's concept of containment. The US went through multiple iterations, such as the new world order (George H. W. Bush), enlargement and engagement (Clinton), the war on terror (George W. Bush), and America first (Trump). Only the Obama presidency abstained from putting up a bumper sticker, although it too popularized the concept of the pivot to Asia, or rebalancing, while its critics accused it of "leading from behind." On the surface, the net effect of this three-decade-long search for a new grand strategy has been murkiness, with no clear successor to emerge for Kennan.[3] But in actual practice,

Table 7.1 Demographic capabilities, great powers, 1990–2019, in millions of people

Great Power	1990	2000	2010	2020	Rank (current)
United States	247	278	312	332	2
China	1,155	1,277	1,359	1,394	1
Russia/ USSR	281	146	143	141	3
Japan	123	126	127	125	4
Germany	63 (79 after unification)	82	83	80	5
France	56	59	63	67	6
Great Britain	57	58	62	65	7

Table 7.2 Military spending, great powers, 1989–2019, in billions of dollars

Great Power	1989/1990	2000	2010	2018	Rank (current)
United States	289	280	698	649	1
China	12	23	119	250	2
Russia	238	43	58	61	4
Japan	29	37	54	46	7
Germany	35	33	45	49	6
France	36	40	59	63	3
Great Britain	34	36	59	50	5

Table 7.3 GDP, great powers, 1989–2019, in billions of dollars

Great Power	1990	2000	2010	2018	Rank (current)
US	5,979	10,284	14,964	21,482	1
China	360	1,214	5,812	14,172	2
Russia/USSR	2,659	259	1,638	1,649	7
Japan	3,140	4,887	5,793	5,220	3
Germany	1,592	2,594	3,309	4,117	4
France	1,278	1,442	2,560	2,844	5
Great Britain	1,093	1,327	2,246	2,809	6

the US has been following a distinct grand strategy for the past 30 years, which, for all intents and purposes, amounts to primacy.

As shown in Chapter Four, primacy is a grand strategy of status quo plus, suitable for a dominant state. The number one consolidates further its advantage over the other great powers, without, however, upsetting the status quo by seeking to achieve hegemony. This is done by the number one maintaining forces in excess of the combined capabilities of the next-in-rank powers; and by gaining control over resource-rich areas and strategic chokepoints, enabling both defense and attack from a privileged position of strength. In effect, the determining factor for the US grand strategy, since the end of the Cold War, has been unipolarity: the presence in the system of only one superpower, or, in other words, a power con-figuration in which one state both "excels in all the component elements of state capability" and has significantly more capabilities than the other great powers.[4]

Already by 1991, it was clear that the US faced no realistic peer competitor. Its dominance was not confined to a single dimension, but was at once military, economic, technological, and diplomatic. The US had inherited an unmatched arsenal from its Cold War build-up, which conferred it not only global power projection capability, but also provided it with a technological edge relative to any would-be competitor. In particular, America's lightning-fast victory over Saddam Hussein's army in the Gulf War in a matter of little more than a month at the cost of only 458 casualties (the anticipated figures prior to engaging in combat the fourth-largest military in the world had been estimated at 20,000) was impres-sive enough to prompt talk of a revolution in military affairs, de facto placing the US in a category of its own. Simultaneously, the US economy experienced a revival of fortunes in the 1990s, aided by the digital revolution, reduction in Cold War military expenses, and the difficulties experienced by some of the chief com-petitors—Germany had to integrate economically the former GDR, while Japan was trapped by a decade-long stagnation. Finally, every other great power was either an American ally (Japan, the Western European powers), or still too weak to mount resistance against Washington (China, Russia). The net result was the advent of America's unipolar moment. As Charles Krauthammer put it, "the true geopolitical structure of the post-Cold War world, brought sharply into focus by the gulf crisis, [is] a single pole of world power that consists of the United States at the apex of the industrial West." Paul Kennedy concluded more dramatically: "nothing has ever existed like this disparity of power; nothing. I have returned to all of the comparative defense spending and military personnel statistics over the past 500 years that I compiled in *The Rise and Fall of the Great Powers*, and no other nation comes close." Nor was this a reality that decision-makers, whether American or foreign, failed to grasp. Every single State of the Union address from 1990 to the present contains a nearly compulsory reference to the presence or the need for US global leadership; and no international actor was shown more mani-fest attention and deference by other states, nor exercised control over the man-agement of the international system than the US has in the past three decades.[5]

This means that in practice, the US has moderately improved upon the status quo, by seeking to make its dominating position in key regions of the international system unassailable, specifically Europe, East Asia, and the Middle East. These three regions are currently the most significant areas in the international system because they contain the bulk of the world's advanced industry, energy resources, and population, and also because they are home to the other great powers. As long as the US conserves the number one spot in each of these regions, it remains dominant overall.[6] Consequently, since 1989, the US has resorted to two tactics. First, despite initially reducing its forces from Cold War levels, the US has main-tained from the end of the 1990s to the early 2000s a level of military spending that has consistently accounted for in between 43% and 46% of the world's total military spending, and in excess of 60% of the military budget of the other great powers. This remains the case today, despite budget cuts: the 2018 US defense budget of $649 billion is about level with the $ 647.2 billion representing the combined spending of the next eight spenders on the list, which comprises five of the other six great powers: China, Saudi Arabia, India, France, Russia, the UK, Japan, and South Korea. Thus, in theory, every imaginable combination of the forces of the other great powers may not be sufficient in order to defeat Washington on the battlefield. Second, since the Cold War ended, the US has expanded its influence further in order to better defend and stabilize its presence in the three key regions, particularly through the conclusions of alliances and security partnerships with local states and through basing accords for American troops. In other words, since Washington's allies and partners are the essential assets for primacy, it is essential that they are made secure by acquiring further allies and partners, a modern reproduction of Britain's turbulent frontier, this time under the formula of the alliances going out of their previous area, or out of business. As Henry Kissinger put it regarding the expansion of NATO, "NATO cannot long survive if the borders it protects are not threatened while it refuses to protect the borders of adjoining countries that do feel threatened." Accordingly, since 1990, in addition to its Cold War system of alliances, the US has gradually acquired a host of new security interests in: a) Eastern Europe and the Balkans, with 13 new members joining NATO; b) the Gulf region (mainly the Gulf Cooperation Council countries, Iraq and Yemen); and c) South and Southeast Asia (Afghanistan, Vietnam, Singapore, Indonesia, and India).[7]

It is important to emphasize that primacy has never been explicitly referenced as a bumper sticker, although there was a good deal of rhetoric on leadership and the indispensable nation, which essentially means the same thing. How then does one know that the US follows primacy? A strong clue, apart from the actions depicted in the above, which are fitting the primacy model, is that the US has either expressly rejected or failed to consider alternative strategies, such as isolationism, hegemony, and concert. In other words, the US has chosen primacy over its competitors. Thus, the available evidence suggests that primacy is the result of decision-making in the George H. W. Bush and first Clinton administrations.

Back in the early 1990s, academics either expressed concerns or could not resist putting forward projects about the US returning to isolationism. Since the threat posed by the Soviet Union was gone, the US was going to disengage from Europe and East Asia, leaving these regions to chart their own path. However, US decision-makers never even thought about, let alone seriously debated, the possibility of pulling back to fortress America.[8] Far more serious consideration was given to the alternative of hegemony. In 1992, the Pentagon, headed by Dick Cheney, proposed a Defense Planning Guidance (DPG), which argued that the US should take active steps to prevent the reemergence of a peer competitor, as formidable as the Soviet Union had been. Hence, in practice, the DPG called for would-be rivals, even those that figured among America's allies, such as Germany and Japan, to be prevented from catching up to it either economically or militarily. The DPG, therefore, went much farther than status quo plus, which merely sought to make the US position unassailable. Instead, it opened the vista of preventive action, including sanctions, and even military action, to slap down competitors. In this line of thinking, allies were not essential assets in enabling the US to stay number one, by enabling and strengthening its presence in key regions, as primacy argued. Instead, they were thought of as potential threats-in-the-making. However, the George H. W. Bush administration unambiguously shelved the DPG—as Brent Scowcroft, the National Security Advisor reminisced: "that was just nutty [...] it didn't go anywhere further. It was never formally reviewed."[9]

Perhaps the most serious alternate competitor to primacy was concert, the grand strategy in which a great power committee would cooperate to ensure international order and security. Concert, stemming from the modus operandi of the international coalition during the Gulf Crisis of 1990, was strongly considered by the George H. W. Bush administration. Saddam Hussein's seizure of Kuwait provoked an unprecedented unified international response from the great powers, which acted together to condemn, impose nearly universal sanctions, and, ultimately, authorize the use of force against Iraq. Therefore, it was hoped that this cooperative pattern of dealing with a crisis by putting together irresistible international coalitions could be reproduced in the future. As Bush and Scowcroft argued,

> our foundation was the premise that the United States henceforth would be obliged to lead the world community to an unprecedented degree, as demonstrated by the Iraqi crisis, and that we should attempt to pursue our national interests, wherever possible, within the framework of concert with our friends and the international community.[10]

But, as subsequent US decision-makers eventually found out, particularly in trying to end the civil war in Bosnia (1993–1995), the concert could not be maintained. The Gulf Crisis of 1990 had been one of a kind, in that Iraq had violated the taboo of conquest, in place since 1945, by annexing Kuwait, a fellow Arab state, alienating in the process every great and regional power. It also occurred at a moment when Moscow, Iraq's erstwhile patron, was at its nadir, and a post-Tiananmen China sought to repair ties with the US.[11] In different contexts, however, the divergence of the interests between great powers was bound to re-emerge. Furthermore, concert, being a cooperative grand strategy, imposes limitations on the strongest powers, which are constrained to wait for the approval of the lesser ones, and which, in case they fail to secure this approval, are confined to inaction. There was after all a reason why concert was particularly favored by a weaker power, such as Metternich's Austria, whose goal was to tie down the powerhouses of the day, Britain and Russia. But in the case of the US, which itself was the strongest power, a concert would not have served if Washington could not get the other great powers to rally behind its preferred position. This proved repeatedly the case in the post-Cold War, notably in the contexts of the former Yugoslavia and of Iraq. Hence, the pattern that eventually emerged was that, having been rebuked by some of the other great powers, the US refused to be prevented from acting. Thus, it reserved the right to address matters on its own, if need be through the use of force, as was the case in the NATO campaign over the Kosovo conflict in the spring of 1999, and the invasion of Iraq in 2003, neither of which was authorized by the other great powers in the form of a UN Security Council resolution. In other words, if concert advised compulsory multilateralism, the strategy that the US followed allowed for multilateralism only for as long as other powers endorsed its position. In case, nevertheless, that they refused to follow the course chartered by Washington, the latter reverted to unilateralism. As the Clinton team found out the hard way in the context of Bosnia, for as long as the US persisted in asking the other powers for authorizing military action against the Bosnian Serbs in order to coerce them to offer concessions, the conflict raged on. But once the US switched from consultation to a line of informing the other powers about the action to be taken, bombing went ahead, and a seemingly intractable conflict was solved in less than one month. As Derek Chollet and James Goldgeier write, the US decided that it would no longer be bound "by concerns over niceties or allied consensus," and would instead go ahead with a take it or leave it strategy motivated by "the desire to get things done." From Bosnia onwards, the US "acted first, and consulted later," effectively giving up on concert for good.[12]

If primacy is the current grand strategy of the US, will it remain so in the future? For the past 12-odd years, a heated academic debate on which grand strategy the US should follow has taken place between two schools of thought: restraint and deep engagement. Restraint essentially supports the adoption of a grand strategy similar to semi-detachment, in that the great power practicing it remains aloof from overseas undertakings, except for circumstances when another great power threatens to achieve hegemony in a key region.[13] At those times, the great power intervenes by helping the weaker powers in the threatened region to redress the balance, which is why restraint is also known as offshore balancing. But once the danger of hegemony passed and the would-be hegemon is cut down to size, the great power returns to its uninvolved stance by disengaging from the region. Consequently, advocates of restraint argue that the US should use force sparingly and solely under very specific circumstances: either in self-defense, or, as a last resort, after the local great powers have tried and failed to preserve the balance on their own. Furthermore, they contend that the US should maintain a limited number of international commitments, and that whatever alliances and partnerships exist should not be permanent, but maintained only for so long as there is a threat of hegemony that local actors cannot handle. Regarding restraint, permanent allies are actual burdens, and supporting them is akin to providing welfare for the rich: since allies can rely on Washington to fight their battles in their stead, they have no incentives to pay for their own defense or assume responsibility for their own foreign policy, preferring to enjoy an easy prosperity provided by American forces. Hence, restraint advises a major reduction in existing US military spending and international commitments, resulting in the withdrawal of US forces around the world, the closing of American bases, and the termination of US formal and informal alliances in the Middle East and Europe.[14] The benefits of this minimalist stance would be threefold. First, the US would spend less on its military—it currently spends around 4.65% of its GDP on its armed forces, of which 15% is marked for supporting the 200,000 soldiers deployed overseas. But if these forces are pulled out, the US would free these resources for use on its own socio-economic problems back home, while still maintaining adequate forces for self-defense and for making a difference in case a would-be hegemon shows up. Second, the US would eliminate the risk of imperial overstretch by scaling down its goals to a level where they can be easily achieved with existing resources. Third, restraint would also reduce or nullify the pushback from other actors, both states and non-states, against the US, in the form of balancing against its perceived hegemony, insurgency, terrorism, or anti-Americanism. For its supporters, restraint would not have adverse repercussions for US or world security. Great powers in each region will step up their own military capabilities to balance each other; and, at any rate, if they prove unable to do so, the US can always come to the rescue.[15]

By contrast, deep engagement supports keeping current US grand strategy largely unchanged. Its partisans stick with the view that alliances and partnerships abroad are assets, not liabilities, and that, by following restraint's advice, pulling out would result in a world far less congenial to American interests and values. Since the security and prosperity of other states help preserve America's own, deep engagement considers it acceptable to employ force not just in emergencies, but also so as to defend US allies and partners, and to maintain international order wherever challenges emerge. This agenda would be impossible to enact without conserving a sizable forward military deployment overseas. Moreover, for deep engagement, the US should follow the opposite course of action of aloofness: that is to say, it should strengthen existing alliances, while seeking to gain even more new allies. Adopting restraint would be a considerable mistake for supporters of deep engagement for several reasons. First, they point out that the US fares better with the devil it knows: the costs of primacy are known, but no one has a clear idea of what the costs or implications of restraint are likely to be, making a change of grand strategy a risky experiment to run. Second, they warn that US withdrawal from the world would be followed by a vacuum of power across several regions, which would generate arms races, crises, and war. With no great power or combination of powers being able to efficiently fulfill the role played by the US, the world would become much less stable, which in turn would end up affecting American security and prosperity. Third, once it pulls out its forces from a given region and it abandons its alliances, the US may find it hard to make a speedy and efficient comeback if an emergency were to occur. By the time the US would be ready to act, it might be too late, with the would-be hegemon having already secured a decisive victory over the local powers, and having added their capabilities to its own. Finally, deep engagement retorts that some of the adverse consequences primacy restraint warns of have never come true: despite the benefit of three decades, no grand coalition of great powers has assembled to balance the US; and imperial overstretch can be compensated by economic benefits arising from the US ability to set up rules and provide the international currency.[16]

A third contender strategy may eventually intervene in the above debate: a revamped version of containment. The argument for containment is that if China continues to rise, it risks catching up to the US in multiple dimensions, not just economically, but also militarily. Therefore, the US would benefit from ensuring it conserves its advantage relative to China, primarily by preventing Beijing from supplanting its influence in East Asia. Early indications of a shift toward a renewed containment have been the above-mentioned pivot to Asia, which notably recommended the deployment to the region of more than half of the existing US naval capabilities, and the designation of China, alongside Russia, as "revisionist powers" and "competitors" in the 2017 National Security Strategy.[17] While containment is still a nascent strategy contender, its formulation is likely to pose significant challenges. First, whether containment should be at all considered depends on whether the US is able and believes itself able to preserve the gap between itself and China in key dimensions. In effect, the more unbridgeable this gap, the less there is a reason to do away with primacy. Conversely, the more China seems to be catching up, the more the window on the US switching to containment narrows—any such shift should happen while the US is still ahead, because, if it has already fallen behind in multiple dimensions, it would already be too late. The precise threshold for making this call on switching from primacy to containment is likely to be up for a lot of future discussions, as is the matter of whether the US is in decline or still ahead relative to China.[18] Second, while containment is a grand strategy with a proven record and a clear set of principles, it should be adapted to the adversary at hand. China has very different strengths and weaknesses than the Soviet Union did. It is not driven by a messianic vision; its economy is far stronger than that of the Kremlin; its population is triple that of the US; and, at the same time, it is more geopolitically vulnerable than the Soviet Union, particularly to a naval blockade of its industrial, trade-dependent south. Third, the containment of China would prompt a reevaluation of existing alignments. Consequently, the US may need, as Great Britain did, to disengage or reduce its footprint in other regions, even in the Middle East and Europe, in order to bolster its position in East Asia and adjacent regions; and may have to gain the support, or at least the neutrality of a number of key countries, valuable either for their resources or their positioning, primarily Japan and India, but also, as discussed below, Russia, Pakistan, Iran, and Turkey.

China

Chapter Three discussed at length China's grand strategy for most of the post-Cold War, which continued Deng Xiaoping's line of reassurance and biding one's time. But this discussion does not account for the recent evolution of China's grand strategy. Over the last seven years, China has undergone a change of grand strategy, following the accession to power of Xi Jinping as General Secretary of the Communist Party and President of China in 2013. Xi is the first post-Cold War Chinese leader not to have been hand-picked for succession by Deng. Since assuming power, he has become perhaps the most powerful Communist leader since Mao Zedong. Notably, he has removed the ten-year service limit for the position of President, which had been in effect since the days of Deng, thereby creating the potential of a lifetime mandate. Xi is thus a consequential leader domestically, and intends to achieve a similar significant legacy on the international scene. This ambition materialized in Xi's initiative to supplant Deng's grand strategy with the vision of the China Dream (*zhongguo meng*).[19]

To be sure, there are additional reasons why China may want to move away from reassurance. The year 2008 was a watershed for China. On the one hand, China discovered a renewed confidence, due to Beijing's hosting of the Summer Olympics, preparations for which were organized by Xi. From the Chinese perspective, the Olympics were a celebration party for China's economic achievements, as well as a showcase for the recognition by the international community of its new importance as a great power. On the other hand, only months after, the US was caught by surprise by a severe financial recession. While the US and most Western economies shrank, China was the first to recover, achieving double figures growth in 2010. Therefore, this was seen by many in Beijing as the point when momentum shifted away from a declining US, and toward an emergent China. Another significant consideration is that, by the end of the 2000s, it was becoming impossible to maintain the charade that China was a backward, run-of-the-mill developing state, and, in so doing, continuing to keep a low-profile. In 2010, China officially became the second largest economy in the world by all methods of measuring GDP, and was also the second largest state in terms of military spending. A final impulse toward a change of grand strategy came from nationalism. Deng himself had encouraged the spread of nationalism in Chinese education and culture in the 1990s as an antidote to the calls for liberalism. But nationalism took a life of its own, imposing constraints on what the Communist Party could say or do. Hence, a strategy of concessions, calm, and measured rhetoric became increasingly stifling as more voices, especially in the military, called for a China that could stand up, or even supplant, the US as the dominant power.[20] To put it differently, since there was a demand for a change of grand strategy, Xi provided the leadership to implement it.

China's dream of national rejuvenation is fundamentally about restoring China to greatness. Once the dominant power of the East Asian system for centuries at a time, China is heir to a long and proud imperial tradition, in which it embodies the de facto center of the world (hence the Middle Kingdom). China is ideally suited due to its size as well as cultural and technological achievements for the role of leading the rest of the world. Following the Confucianist worldview, imperial China portrayed its role as a benevolent suzerain, exercising a wise and moderate role, but demanding in return strict obedience and formal acknowledgment of its overall superiority. Yet China's claim to systemic leadership was usurped by the Western powers and Japan in the so-called "century of shame," (*bainian guochi*) stretching from the Opium Wars to the Korean War. In this interval, China was dismembered, subjected to colonization and occupation, and forced to accept the principle of extraterritoriality, admitting the supremacy of foreign laws over Chinese rule in its own territory. From the Chinese perspective, China suffered not only demotion, but also national humiliation, by being reduced to an international rank and extended a treatment far below its ambitions.[21]

What the China Dream seeks to achieve is, therefore, to vindicate past wrongs and return to normality, fairness, and order by restoring China to its rightful place at the top. In practical terms, this is a claim for renewed regional preeminence in East Asia (comprising both North East Asia, meaning the two Koreas, Taiwan, and Japan; and Southeast Asia, meaning the ASEAN countries); and for a favorable environment in terms of security and economic opportunities in the areas beyond it: Central Asia, South Asia, and the Middle East. To quote Yan Xuetong, "the Chinese regard their rise as regaining China's lost international status rather than as obtaining something new. This psychological feeling results in the Chinese being continuously dissatisfied with their economic achievements until China resumes its superpower status." Or as Ye Zicheng argues: "if China does not become a world power, the rejuvenation of the Chinese nation will be incomplete. Only when it becomes a world power can we say that the total rejuvenation of the Chinese nation has been achieved." Immediately after becoming General Secretary, Xi reportedly brought the other Chinese leaders to the National Museum for a solemn pledge of restoring China as a great power of the first order economically, militarily, and culturally by 2049, hence a century after the Communist Party took power.[22]

Consequently, Deng's injunction about maintaining a low profile has been replaced with Xi's advice to strive for achievement (*fen fa you wei*) and take greater initiative. China has started referring to itself as *daiguo*, a big country or great power, and argued that it should become a rich and powerful country, and not just a rule taker, but also a rule maker, thus ensuring that its interests and values are reflected in the international system management institutions. It has clamored for "a new type of relationship between major powers," , thereby de facto claiming full equality with Washington. Even more significantly, China realizes full well that its current center of gravity lies in its south, where the special economic zones and Shanghai lie. The problem is that this coastal region is particularly vulnerable from the sea, especially considering the superior capabilities of the US navy, either to attack, or, more likely, to put into effect an economic blockade. Hence, if China is cut off from foreign raw materials and energy supplies, and it is no longer able to rely on exports to generate capital, it can be starved into submission.[23] Accordingly, the whole crux of the China Dream strategy comes down on how to prevent this scenario from coming into being by utilizing a two-pronged approach.

One prong has been the expansion into the South China Sea. China has long claimed an exclusive zone, extending from its territory and covering about 80% of the reefs and islets of the Paracel and the Spratly Islands, the so-called nine-dash line. But it is only in the last few years that it has designated the area, swaths of which are also claimed by other countries such as the Philippines, Taiwan, or Vietnam, as a core national interest on par with Tibet or Taiwan. In the space of only eight months, from December 2013 to August 2014, China expanded in the area by dredging and reclaiming more reefs than all the claimant states in the area have done over 40 years. China has been busy fortifying these positions, with the goals of claiming ownership over the area, and, more importantly, of keeping the US warships at bay as far as possible from the Chinese shores. This creeping advance is coupled with a major naval build-up, which has transformed China into the owner of the second largest navy in the world. As the Chinese Defense White Paper of 2015 argues, "the traditional mentality that land outweighs sea must be abandoned, and great importance has to be attached to managing the seas and oceans and protecting maritime rights and interests." Hence, China is opting for an aggressive defense, moving from an "offshore defense" to an "open-seas protection" strategy, thus forming a maritime buffer for its southern provinces.[24]

Yet, concomitantly, China also pursued a second prong in its strategy: the so-called Belt and Road Initiative (BRI). The project officially stands for China building or financing a modern-day equivalent to the ancient Silk Road. This is done first and foremost through the construction of a complex network of railroads, highways, harbors, pipelines, and fiber-optic cables. One part of the project, the Belt, would link by land China, Central Asia, West Asia and the Middle East, and Europe. It would originate in Xian, going through Almaty, Samarkand, Dushanbe, Tehran, Istanbul, and Moscow and ending up in Hamburg. Meanwhile, the other part or the Road, will connect by sea China with Southeast Asia, South Asia, the Middle East, and East Africa, starting in Guangzhou, going through Jakarta, Singapore, Kolkata, Nairobi, Athens, Venice and ending up in Rotterdam.[25] The BRI, which has received the highest degree of support from Xi, and to which China has already committed 1 trillion dollars, has several goals. First, it circumvents American encirclement by minimizing the impact of a potential blockade. While the US relies on its naval superiority and on its superior position in the Pacific, especially its alliances with liberal democratic Japan and Australia, it has considerably less leeway with the countries to the west of China, many of which are ruled by authoritarian leaders, and show hostility or ambivalence toward American interests (Russia, Iran, Pakistan, Kazakhstan). Therefore, if supplies by sea in the east and south are endangered, China can turn to bringing them overland by pipelines, or over sea lanes to its west. It is in fact estimated that China may import from Russia, Iran, and Central Asia about two-thirds of its total requirements of natural gas, and about a fourth of its oil.[26] Second, as the US pivots east toward East Asia, gathering allies and increasing its naval capabilities in the area; China pivots in precisely the opposite direction: west, toward Central Asia and the Middle East. Thus, it avoids clashing directly with the US, while advancing in an unexpected direction.[27] Third, the BRI also undermines an eventual American diplomatic containment of China. The US is greatly dependent on its allies, but if many of these allies are also heavily involved economically and culturally with China, which is done not merely through trade and infrastructure development, but also through investment, communications, and education, they may decide to either jump ship in the Chinese camp, or restrict its cooperation with the US, thus debilitating any future effort at containment. Presently, China has only one official alliance treaty with North Korea, while the US has alliance commitments with no fewer than 70 countries. But, on the other hand, China has managed to, so far, attract more than 60 countries to take part in the BRI, including a number of US formal allies, such as Pakistan, Saudi Arabia, Poland, Czechoslovakia, Turkey, Greece, Italy, and Germany. If the area comprising the BRI is taken as a whole, it would account for more than half of the population of the world (4.4 billion) and for a quarter of its GDP at 21 trillion dollars. This means that if Eurasia ever becomes a political reality, a space under Chinese leadership, it would effectively surpass the combined resources of the US and of its other allies. Fourth, the BRI plays to Chinese strengths, by putting to good use surpluses of construction as well as its expertise in building up China's infrastructure; by providing capital to countries in dire need of outside investment; and by adopting an inclusive attitude, in which China is interested in business, not politics, and therefore does not condition economic support, as is the case with the US, to progress in human rights and democracy.[28]

The China Dream strategy, however, also poses two risks that Chinese decision-makers will have to contend with in the future.[29] The first such risk is gigantism. Being the leader in Eurasia is undoubtedly tempting, but China's control of the region does not depend on its control over this extensive real estate, which is likely to be prohibitively expensive, and some of whose denizens are at odds with one another (India and Pakistan, Russia and Poland, Saudi Arabia and Iran). Furthermore, it would be a grave mistake to assume that economic connections may overweigh security ties. As long as China is not able to provide its BRI protégés with security, they would have to strike compromises with the US, thereby remaining part of the American camp. Besides, the more inclusive the BRI gets, the more countries China is asked to sponsor and protect. A more minimalist orientation, aiming only at securing alliances and partnerships with the most powerful and strategically located states, primarily Russia, Pakistan, India, Iran, and Turkey, would, therefore, achieve China's key objective of ensuring resources and avoiding isolation at a lesser cost. In fact, without these countries' support, the BRI would be impractical even if embraced by everyone else; while, with them on its side, the rest of the countries in South Asia, Central Asia, and the Middle East, would either fall into line or become irrelevant.

The second risk posed by the China dream is that it essentially represents a resurgent declining power grand strategy, not that of a rising power. In other words, if China aspires to achieve the kind of dominance enjoyed by Imperial China, it may set the bar too high for itself, and, in the process, create unnecessary suspicion and hostility. It should be remembered that Imperial China was perhaps the closest thing to a hegemon, whose only possible counterpart was the Roman Empire, reigning over a number of subordinate states. Such a position of complete dominance contradicts the very rules of the modern great power system, based on the existence of independent sovereign states. It would be easy for competitor powers to exaggerate, spread, or take advantage of misgivings among both China's neighbors and among great and regional powers that seek to play an important part in Eurasia.[30] Accordingly, China may fare better, if, instead of trying to recreate the golden times of the Han or Tang dynasties, it were to focus its energies on becoming a world power, on the model of the US, or, even better, if it were to declare itself fully satisfied with what it has achieved so far, declare its rise as completed as the number two power in the system, and patiently wait while American power ebbs away. By defining itself as a rising power set out to obtain something new instead of being busy recapturing its ancient brilliance, China may choose from the rising power grand strategy menu. In particular, it is Bismarck's divide and conquer/tertius gaudens grand strategy that may prove the most useful for its situation. This does not mean, as in Bismarck's days, creating a web of crisscrossing alliances, but rather, always being there as an alternative for states dissatisfied with US leadership or looking to enhance their bargaining position vis-à-vis Washington, thereby essentially securing gains from the inevitable internal disputes of the competing alliance. If, as in Bismarck's days, China wrests away the support of sufficient key actors from the US, it will always enjoy superiority in a possible contest. A complementary tactic may well be to speculate containment's major weakness: the psychological need to secure the appearance of victory in contexts that are not really worth fighting over. By issuing challenges in secondary contexts, China may be able to stretch the US thin over a wide theater of operations and gradually deplete its resources—basically the Cold War playbook of Vietnam and of Afghanistan, where one side successfully lures the other into quagmires. For its part, it is in the interest of the US not to play China's game. As long as the decisive great and regional powers are on its side, it is the one enjoying the upper hand in Bismarck's three-out-of-five calculation, no matter what reverses are suffered in secondary quarters. Washington should therefore do well to remember Kennan's advice and be particularly cautious of any impulse to mount intervention in any areas that are not significant either in terms of resources or geopolitical location.

Russia

"The collapse of the Soviet Union was the greatest geopolitical catastrophe of the twentieth century," Vladimir Putin told his countrymen in April 2005. For emphasis, he reiterated this view 13 years later, when asked which particular historical event he would like to change. This is not Communist nostalgia.[31] What Putin is lamenting is rather the decline of Russia as a great power, from a superpower to a rank-and-file member of the great power club. In effect, seen from this perspective, the loss of power and status Russia suffered in 1991 is staggering. Its territory was diminished by more than 2 million square miles (5 million square kilometers), which is the rough equivalent of half of Europe; its population was basically halved from 281 million to 140 million, and its GDP plunged by a factor of ten. The consequence of this sudden collapse was not, as the architect of the move, Boris Yeltsin, had hoped, the emergence of a leaner, more efficient Russia that could escape the burden of supporting the other 14 Soviet republics, and reshape itself on the model of the West. Instead, the Russian economy went into tailspin, experiencing negative growth for eight consecutive years, as a consequence of a combination of falling oil prices, disruption of the Soviet supply-and-demand chain, which now had broken down into separate pieces, each in a different country, drastic liberalization (the so-called shock therapy), corruption, and ever-growing foreign debt. By 1998, Russia had to default on debt-payments: it was officially bankrupt.

By that time, the Russian state was barely hanging on in one piece. In 1993, Yeltsin laid siege to his own nationalist-controlled Parliament. In 1994, he launched a military offensive against Chechnya, a small breakaway republic in the North Caucasus area, only for the Russian troops to suffer humiliating defeat. Russia then had to conclude an agreement with the separatists, which Putin later called "a capitulation," that left Chechnya practically independent. In 1996, to secure his re-election Yeltsin rented Russia's most valuable commodities, its raw materials and energy resources, to a group of nouveau-riches, the so-called oligarchs, who, in return, funded his campaign. The same year, he suffered a heart attack and had to undertake surgery for seven hours. With the political succession very much insecure, an ailing and increasingly inebriated president, and a slew of powerful oligarchs and local governors in control, Russia was the weakest it had ever been since its Civil War, prompting the description of this interval as a renewed "time of troubles," in reference to the seventeenth century, when due to internal dissensions, the country ended up under foreign occupation.

It was no wonder that this chronic weakness had consequences in Russia's stance among great powers. Uncertain of the outcome of the power struggle going on in Russia, the West was reluctant to include it in either the EU or NATO. In fact, the best remedy to Western decision-makers seemed to ward off the instability and ethnic conflicts, re-emerging in the aftermath of the collapse of Communist regimes all over Central and Eastern Europe, as well as the Balkans, by expanding Western institutions in these quarters. Accordingly, NATO was first enlarged in 1997, by the inclusion of Poland, the Czech Republic, and Hungary, and promised a further expansion (realized in 2004), which brought in Romania, Bulgaria, Slovakia, Slovenia, and, most importantly, Latvia, Estonia, and Lithuania. This was particularly problematic for the Kremlin: the US effectively was expanding its sphere of influence in the area that had previously been part of its sphere, and, moreover, even in an area that had been actual Soviet (and Russian) territory and which had controlled, from the times of Peter the Great, Russia's access to the Baltic Sea. Kennan, now in his nineties, warned vainly that the move constituted "the most fateful error of American foreign policy in the entire post-Cold War era," because it would impact adversely Russia's security and status. Essentially, he warned, this would be interpreted as the US taking advantage of Russia's near-paralysis and would lead to lasting resentment. This seething resentment was then compounded by the 1999 bombing of Yugoslavia over the crisis developing in Kosovo. The crisis was seen as a major humiliation by Russia, on account of the facts that the Kremlin was acting as a protector of the Serbs; that NATO had shown its "true colors" as the perpetrator of what Russia saw as naked aggression; that its veto in the Security Council, one of the last remaining instruments that Russia had left at its disposal to assert its great power status, had been disregarded; that Yeltsin's warnings of a potential nuclear war were dismissed as bluff; and that its efforts at sending assistance to the Serbs floundered on the resistance of would-be NATO members Romania and Bulgaria, which blocked its access, and on the cost of logistics. In other words, Russia had been exposed for all to see as a state that could barely intervene in its neighborhood, and as a *quantité négligeable* that other great powers did not bother to seriously consult or take into account in their decisions.[32]

The key question for Russia became therefore how to arrest and, eventually, revert this decline. Unsurprisingly, its new grand strategy under the leadership of Vladimir Putin focused on reinstating Russia as a great power. In August 1999, Yeltsin designated Putin as Prime Minister, and in December of the same year he announced his resignation in his favor. Only days before this announcement, Putin unveiled his vision in a 5,000-word manifesto. Russia, concluded Putin, while coolly reviewing in exact figures just how far behind the state had fallen by comparison to the US and China, was facing one of the most difficult periods in its history. For the first time in the past two or three centuries, it was facing the real threat of sliding into the second, and maybe even the third echelons of states, hence risking dropping out of the great power club. This, implied Putin, was unacceptable: the very idea of Russia was based on the state's greatness, or in other words, being a great power was an essential part of Russian identity. Therefore, Putin pledged that

> Russia was and will remain a great power. It is preconditioned by the inseparable characteristics of its geopolitical, economic and cultural existence. They determined the mentality of Russians and the policy of the government throughout the history of Russia and they cannot but do so at present.

As a result, he announced as his key objective the restoration of Russia as a leader, not only in areas such as economy, technology, and the well-being of its population, but also in ensuring its security and in upholding its interests in the international arena. Russia was not going to be a pushover anymore.[33]

A large part of Putin's agenda consisted of economic, political, legal, administrative, and ideological reforms and had to do with the internal strengthening of the government of Russia. By 2008 and the onset of the world recession, Russia had experienced eight years of continual economic growth, in fact, at an average of 7%, one of the highest rates in the world, fueled in large part by the climbing price of oil, which went up fourfold between 2004 and 2008. Accordingly, Russia could afford to pay out most of its international debt, reign in the oligarchs, and repress the separatists in Chechnya through a successful military campaign and follow-up peace maintaining operation. In 2008, it began a comprehensive rehaul of its military, which in the ensuing decade resulted in more efficient structure, training, and in the eventual equipping of 70% of its armed forces by 2020 with new weapons (under ten years old). Finally, Putin put in place, if not a democratic state, then at least a functional and orderly one. This normalization of internal conditions enabled Russia to reassert itself internationally.[34]

The fulcrum of Russia making a comeback as a great power is the so-called Near Abroad, the space occupied by the successor states of the Soviet Union, minus the Baltic Republics that have joined NATO and the EU. The logic is that by adding the demographic, natural, and economic resources of these countries to its own, Russia would be able to obtain once again enough critical mass to be considered a world power. Furthermore, the Near Abroad states suffer the same structural and political impediments as Russia did in the 1990s, without enjoying any of its advantages; as a result, they have remained militarily vulnerable, and politically and socially brittle. Russia is perfectly positioned on their doorstep to exercise influence, either through military intimidation or economic blackmail. Besides, Russia can also count on the support of a large number of ethnic Russians, an estimated number of 20 million, most of them concentrated in Kazakhstan and the Ukraine, but also with percentages of around a quarter of the population located in Estonia and Latvia. Some of the native Russian or Russian-speaking minorities occupy border areas close to Russia: in the case of Georgia, in South Ossetia and Abkhazia; in that of Moldova, in Transnistria; and in that of Ukraine, in Donetsk and Luhansk. These minorities can be used either to exercise pressure on their host state, or to justify intervention on their behalf by Moscow. It is important to emphasize that Russia's objective in the Near Abroad is not the reconstitution of the Soviet Union or of a Tsarist-like empire, which would impose inordinate costs of maintaining unity in the face of divergent nationalist movements from each country. One of the rationales behind the dissolution of the Soviet Union was precisely avoiding the sort of nationalist rivalries that led to civil war in Yugoslavia. Thus, Russia is interested instead in setting up its own sphere of influence over the former Soviet space, affecting the foreign policy of the states within, while allowing them autonomy in their domestic interests, as long as they do not touch upon their alignment with the Kremlin.

Russia has followed two types of tactics in making headways in the Near Abroad. In the first place, it has attempted to integrate a number of the former Soviet states under common economic and military institutions under its leadership. Following the collapse of the Soviet Union, Russia had set up a successor organization, the Community of Independent States (CIS), reuniting ten of the previous 15 Soviet republics, with two more signing but not ratifying the treaty, and one (Georgia) exiting it. However, the CIS was too loose in its terms for military cooperation or for setting up a free trade area. As such, it suffered both from initial Russian neglect, as Yeltsin snubbed relations with his neighbors in favor of courting the West, and from the reluctance of existent members to agree to tighter relations. For this reason, Russia under Putin has decided to start the integration process over, by launching several fresh projects, notably the Collective Security Treaty Organization (CSTO), which brings together Russia, Belarus, Kazakhstan, Armenia, Kirgizstan, Tajikistan, and Uzbekistan; and the Eurasian Economic Union, comprising Russia, Belarus, Kazakhstan, Armenia, and Kirgizstan. Basically, while the CSTO is predicated on being a military treaty, based on the principle that an attack against one of the signatories is an attack against all, hence an alternative to NATO, the Eurasian Union is supposed to be an economic alternative to the EU, allowing the free flows of goods, services, and capital. Russia has also stepped up its efforts to set up a de facto merger of Russia and Belarus, which was agreed upon in a 1999 treaty, but has never been fully implemented.[35]

The second tactic employed by Russia has been to exclude Western powers from the Near Abroad, by resisting the extension of membership for the former Soviet states into NATO and the EU, whether by diplomacy, economic coercion, threats, or, as a last resort, by using force. Putin came to power with a newfound pragmatism. Unlike Yeltsin, he refrained from issuing empty threats: for as long as Russia was weak, he found that it was better to put a good face on Russia's setbacks, dismissing the second expansion of NATO in 2004, which comprised the Baltics, and which was, after all, undertaken, during his watch, as a "mistake." From 2001 to 2004, Putin even dangled in front of the Western leaders the tantalizing prospect that Russia might be interested in joining itself NATO and the EU. This offer might well have been genuine, or a move designed to create dissension among US and Europe and various Europe countries, or, to weaken both institutions from within once Russia was admitted. Regardless, the initiative was never followed through by the West. Instead, NATO and the EU opened the prospect of future membership for interested Soviet states at a very vulnerable time, as so-called colored revolutions occurred in several Near Abroad countries: the Rose Revolution in Georgia in 2003, the Orange Revolution in Ukraine in 2004, and the Pink Revolution in Kirgizstan in 2005. These revolutions brought to power not only democratic, but also pro-Western governments, which was seen as a major threat by Moscow, who suspected that these movements were not only encouraged, but also orchestrated by the US, with the likely ultimate goal of a color revolution taking place in Russia itself. As Putin was to state in 2014, after the latest bout of color revolutions in Ukraine:

> we understand what is happening; we understand that these actions were aimed against Ukraine and Russia and against Eurasian integration. ... They are constantly trying to sweep us into a corner because we have an independent position, because we maintain it ... But there is a limit to everything. And with Ukraine, our western partners have crossed the line ... Russia found itself in a position it could not retreat from. If you compress the spring all the way to its limit, it will snap back hard.[36]

Consequently, once strengthened, Russia adopted a far more confrontational approach toward the US and its allies. In February 2007, at the annual Security Conference in Munich, Putin denounced unipolarity as a system of "one master, one sovereign," and the US as a state that has "overstepped its national borders in every way. This is visible in the economic, political, cultural, and educational policies it imposes on other nations. Well, who likes this? Who is happy about this?" In April 2008, at the NATO summit in Bucharest, Romania, in which it was a guest, Russia succeeded in maneuvering so as to avoid the enlargement of the organization to Ukraine and Georgia, although the summit left open the possibility of them joining at a later date. Barely months after the summit, the Georgian government took military action against separatists in South Ossetia. The mistake of the government in Tbilisi was that, by international agreement, Russia had peacekeepers deployed between the separatists and the Georgian government, which constituted the perfect pretext for Russian intervention ostensibly to safe-guard the safety of its troops, and to respond to what it described as Georgian aggression. The US and Israeli-trained Georgian army collapsed in five days and Russian troops advanced within sight of Tbilisi. The point of the Russian inter-vention was not to coerce Georgia into obedience, but rather to teach a lesson to any Near Abroad republic that would have looked to the West instead of Moscow as a military and economic patron.[37]

This lesson was then repeated in a harsher fashion in the case of Ukraine. Tittering on the edge of bankruptcy and plagued by endemic corruption, Ukraine sought support from both the EU and Russia. Eventually, the government of Viktor Yanukovych snubbed the EU, indicating a tilt toward the Eurasian Union and provoking massive protests in the capital of Kiev in February 2014. Talks between Yanukovych and the opposition, brokered by both the EU and Russia, produced a deal to diffuse the situation and to organize new elections. But pro-tests did not cease; quite the contrary, with the police moving out of the way, the protesters took over the government's offices, driving Yanukovych into exile. Russia's response was swift: five days after the ousting of Yanukovych, Russia used units from its fleet, stationed at Sevastopol, to seize control over the Crimean Peninsula, a key strategic point on the Black Sea. In April of the same year, the Russian-speaking regions of Luhansk and Donetsk declared their independence from Ukraine, igniting an ongoing civil war. As the forces of Kiev appeared to gain the upper hand in July 2014, Russia proceeded to full-scale intervention in August, a fact that it has never openly acknowledged, throwing back the Ukrainian offensive. Like the attack on Georgia, Russia's intervention is fundamentally aimed to keep other great powers out of what it sees as its exclu-sive sphere of influence. Yet this also showcases the current contested situation in the Near Abroad. Russia certainly enjoys the allegiance of at least some of the former components of the Soviet Union, perhaps most importantly of Belarus and Kazakhstan, important for their strategic location and resources, respectively. But this allegiance cannot be taken for granted, as several of Russia's allies, including the ones in Minsk and Astana, have also kept flirting with the US and the EU. Moreover, Russia lacks important pieces of real estate for its sphere of influence to take shape and be consolidated: primarily Ukraine, one of the foundations of Russian power since before Peter the Great. However, unlike the other great powers, Russia has demonstrated it is ready to go to war to secure the position of Ukraine in its sphere. As such, it is currently enjoying the upper hand in the area, but its hold remains tentative.[38]

Russia's determination to gain and maintain a hold in the Near Abroad has resulted in a more assertive stance toward the US and NATO, especially from 2014 onwards. In order to level the playing field against the superpower and its allies, which are superior in both economic and military capabilities, the Kremlin had to both make the most of its existing strengths, and to be nimbler in using its power than its rivals.

Accordingly, Russia employed to great effectiveness its advantage as a large oil and, in particular, natural gas producer. Under Putin, the Russian government gained ownership over the largest oil and gas companies (Gazprom and Rosneft), hence of production, as well of the railways (Russian Railways) and of the pipeline system (Transneft), hence of transportation. It thus could put in place a price system, where states in the Near Abroad that comply with Russia's demands can acquire energy at cheaper rates than the market price, but are denied such preferential treatment if they cross the Kremlin. Since high oil prices are essential to its economy, Russia has maintained a partnership with Saudi Arabia, the number one oil producer and de facto leader of OPEC, since 2016 so as to coordinate production, maintaining prices at a high level in the face of American efforts to reduce them. Finally, the Kremlin has given careful attention to the development of an extensive pipeline system transporting gas toward the West: the Nord Stream, underneath the Baltic Sea, linking Russia with Germany; the Turk Stream, underneath the Black Sea, linking Russia, Turkey, Bulgaria, and Hungary; and a third projected system is to go through Turkey, Greece, and the Adriatic to Italy. These projects not only bypass some of the states that have shown the most opposition to Russia's projected sphere of influence in the Near East (Ukraine, Poland, Romania, Sweden), but also create a connection with powerful players in the EU, such as Germany and Italy, who, as a result, develop a stake in keeping open economic connections to Moscow, and, hence, will suffer the consequences, if sanctions are put in place.[39]

The second asset Russia has deployed to gain leverage over the West has been the nuclear arsenal it had inherited from the Soviet Union. During the 1990s, Yeltsin had consented to remove the multiple independently targeted reentry vehicles (MIRV—essentially multiple warheads) from Russian missiles, which would have resulted in a lower warhead number for Russia than for the US. But the treaty was never implemented. As NATO expanded into Eastern Europe, and its military capabilities withered away, Russia found that holding on to its nuclear missiles was one of the few tools it had left at its disposal to assert its great power status. It thus dropped its previous no-first-use doctrine, which had been put into place in an era when the Soviet Union enjoyed conventional forces, and by 2000, it announced that it reserved the right to use nuclear forces against a conventional attack. However, Russia has grown increasingly concerned by the US pursuit of a missile defense system, which is meant to protect the US and its allies in Europe and East Asia, respectively, against a nuclear missile launch from Iran or North Korea. The system, nevertheless, also reduces the value of the Russian nuclear deterrent—because if the US becomes invulnerable to foreign missiles, it might in turn attack with impunity and, furthermore, because Washington deployed part of it in Eastern Europe, non-coincidentally in the countries wariest of the potential of Russian expansion: Poland and Romania. Russia has thus answered the US withdrawal from the ABM Treaty in 2002 with its own withdrawal from START II. It also responded to the deployment of radars and of anti-ballistic missiles in Eastern Europe by developing hypersonic missiles, which traveling at a speed faster than sound would be able to avoid NATO and US defenses, and by deploying a cruise missile, the SSC-8, in the west (NATO accuses Russia that the missile has a longer range than that allowed by the INF Treaty, while Russia denies that this is the case).[40]

Russia has also scored a number of limited victories over the West by moving where the US refuses to tread, and by finding new uses for its cyber-capabilities. Since 1989, when Gorbachev ended Soviet economic, military, and diplomatic support for clients and allies in both the Third World and Eastern Europe, Russia has been confined to its immediate neighborhood. But in the 2010s, with the US disengaging from Iraq, and reluctant in its aftermath to take on yet another mission in the Middle East, Russia has moved in. In doing so, it took advantage of a remaining lone naval base in Syria, as well as of diplomatic connection to Iran, a long-time customer for its weapons industry. As the US refused to intervene in the Syrian Civil War, Russia stepped in in 2015, allowing the government in Damascus not only to survive, but, for all practical intents, to also win the civil war against domestic insurgents and ISIS. Like the British forays in Europe in the nineteenth century, this is an instance of making a power display in those areas where it is cheapest to bring Russia's forces to bear. As a result, Russia wins status as an effective power, demonstrates its credentials as a would-be world power able to affect events outside its own region, poses as a credible great power patron and alternative to the US, and at the same time avoids running into US opposition. In the aftermath of the intervention in Syria, Russia made a number of additional military and diplomatic forays in Latin America (Venezuela), the Middle East (Libya), and Africa (Mozambique, Central African Republic).[41] Finally, Russia has exposed US vulnerability in an unexpected quarter: its domestic politics. Previous to 2016, the discussion of cybersecurity was confined to uses of cyber warfare for undertaking sabotage of key enemy military or economic systems, or for conducting military or industrial espionage. Russia has found, however, a different use of cyber weapons: not for attacking the opponent's hardware, but rather its software, meaning its domestic politics and morale, by the dissemination of false information (fake news), and by manipulating opinion on social media platforms such as Twitter and Facebook to support or oppose political candidates that could either endorse or resist Russia. This unorthodox tactic is an extension of the new Russian concept of hybrid war put forward by the Russian Chief of the General Staff Valery Gerasimov. As the theory goes, since there is no clear delimitation between war and peace, hostilities can be conducted not just through the use of open or covert military force, but also through a wide range of operations including cyberattacks, economic pressure, humanitarian appeals, and not least propaganda, deception, and disinformation.[42]

It is, however, debatable to what extent Russia's grand strategy of revival as a world power has been successful. Russia is a great power unlike the rank-and-file club members, which justifies its inclusion as a power of the first rank. It still has the largest territory, the third largest population overall and the first one in Europe (double, if one is to also consider the ethnic Russians and Russian-speakers in the Near Abroad, that of second-place Turkey and Germany), albeit aging, the world's largest resources of natural gas and the second-largest of oil, as well as being the co-owner of the world's largest nuclear arsenal and of the likely third strongest military force in the world. Yet, Barack Obama dismissed it as a regional power lashing out at its neighbors out of weakness, and its current GDP ranks outside the world's top ten economies. Even when not under sanctions for its occupation of Crimea and interference in the Ukraine, Russia's economy only managed to fall in the same bracket with Great Britain and France, the lowest performers in the great power club.[43] To a large extent, present Russia resembles its Tsarist incarnation in the second half of the nineteenth century, formidable in terms of natural resources and military force, but reliant on a weak economic and technological foundation. Thus, turning Russia into more than an oil and gas producer and nuclear power, which was Putin's goal in his 1999 manifesto, was never realized. Also, Russia has got the respect and attention of the other great powers, but this may be the wrong kind of attention: rather than managing to reconcile them to its dominance of the Near Abroad, it has managed to alienate them to the point where the US describes the Kremlin as a power competitor similar to China, and the European countries have imposed economic sanctions in conjunction with Washington. Russia should be careful not to imitate the example of Napoleon III, by throwing away the capabilities it has painstakingly reconstituted in a hasty effort to make a comeback. Rather, it might consider borrowing from Stalin's lying-in-wait grand strategy. The more Russia stays aloof of existing alignments, the more it can gather strength, and the more the forming camps would find it necessary to bring it on board. With the US and China likely to be trapped in a rivalry over the control of Asia, Russia can act as a kingmaker, giving the upper hand to the side it chooses to support. Essentially, like Stalin, Russia is in the position of running a continual auction for its allegiance, demanding concessions in the Near Abroad from both Washington and Beijing. However, for this strategy to work, it is necessary that Russia abstains from committing early to either side, keeps open channels to both, and stays away from the kind of excessive international activism that makes it bump against the interests of other great powers at every turn. Russia's actions in Syria and its interference with foreign elections, notably in the US, may be tactically brilliant, but are strategically unnecessary, distracting Russia from what is essential to its fate as a great power: the pursuit of deep-seated economic and technological reforms, and the formation of a sphere of influence over select parts of the Near Abroad, significant either in terms of strategic location or of industrial, natural, and demographic resources.

Conclusion: grand strategy and the strategist

Good science begins with classification. Hence, a rigorous study of grand strategy cannot be attempted without having in place a basic typology of grand strategies. This book has sought to provide such a taxonomy. Doubtless, this can be revised and improved upon, but it is important to take this first step. Accordingly, the book has defined grand strategy; reviewed its fundamental elements; introduced a distinction between the grand strategies of rising, status quo, and declining powers; explained how the three types of great powers, and hence of grand strategies, can be told apart; and then reviewed, with historical examples, the various grand strategies available to each type. Of course, more needs to be done in future investigations. This is not just about including the great powers that have not been mentioned in here, particularly the other status quo current members of the great power club: Japan and Germany, and France and Britain. The investigation can be broadened to pre-1648 grand strategies of empires, whether in Europe, or very importantly, in non-European instances, such as the grand strategies of Imperial China or the Ottoman Empire. The present typology may also be applied to the grand strategies of regional powers, such as contemporary India, Turkey, Brazil, and Iran. This book has sought to offer examples of successful grand strategies, but equal attention should be given to grand strategies that have failed for each power type, and, in particular to the reasons for their failure. New types of grand strategies may need to be added or substituted in the existing scheme in order to account for the conduct of middle powers (Canada, Sweden, Norway), or of the average states in the system. One further step that should be taken is to consider in-depth cases of individual grand strategies from an analytical angle, as opposed to providing a narrative of the strategy. This endeavor should make clearer whether the grand strategy is substantial, instead of a mere intellectual and perhaps imaginary pattern derived from the statements and actions of the decision-makers, and how the elements of goals, means, and challenges combine and interact to constitute the grand strategy.[44]

This latter exercise is not meant to provide a do-it-yourself, foolproof approach to producing grand strategy. But, ultimately, education as a strategist matters. Perhaps the most unexpected finding of the investigation carried out by this book is that it is impossible to tell the story of grand strategy without mentioning, quite frequently as it turns out, the strategist. The strategist may be either the person formulating the strategy, or the person implementing it in practice; in some cases, the strategist plays both roles. The importance of the strategist is unexpected, because grand strategy, as the theory of raison d'état before it, proceeds from the assumption that any rational actor meeting with that respective situation, pursuing similar goals, and disposing of the same means would choose the same course of action. But this assumption of different strategists, subject to the same constraints and incentives, behaving similarly is manifestly not the case. Various strategists arrive at various answers as to what should be done even though their circumstances may be similar. Just imagine that Hua Guofeng had been in charge of China from 1979 onwards instead of Deng Xiaoping, that Caprivi had led Prussia instead of Bismarck in the 1860s, that Frederick II would have chosen to pursue the same bargaining-for-advantage line as his father vis-à-vis Austria instead of seeking to transform Prussia into a great power, or that Trotsky, and not Stalin, had been in charge of the Soviet Union from 1927 on and confronted Hitler in World War II. This is to say that the strategist-in-charge does make a difference.[45] This is not, however, the same difference as in foreign policy, where personality, bias, beliefs, life experiences, and background come together to determine choices at crucial junctures, usually during crises. Rather strategists are differentiated by their skill in assessing correctly the elements of a situation; their sense of opportunity and risk; their pragmatism in facing up to unpleasant realities; their sense of proper timing in acting or refraining from action; their patience in producing results; their elocution in conveying their vision; their power of persuasion, authority, or ability to coerce and to reward other actors for supporting the strategy; and their flexibility on reaching compromise on various aspects of the strategy in order to make it practical. Strategists are not omniscient geniuses able to see farther than everyone else; supermen or superwomen bending history to their iron will; or, as Lawrence Freedman writes, quoting Harry Yarger, being a living embodiment of perfection.

> aware of the past, sensitive to the possibilities of the future, conscious of the danger of bias, alert to ambiguity, alive to chaos, ready to think through consequences of alternative courses of action, and then able to articulate all this with sufficient precision for those who must execute its prescriptions.

Consequently, Freedman goes on to argue that the master strategist is a myth: "there was only so much knowledge that an individual could accumulate, assimilate, and manipulate; only so many potential sequences of events that could be worked through in a system that was full of uncertainty, complexity, and chaos." Freedman is right in so far as the strategist is not in control of all. Indeed, strategists, as the largely successful heroes of this book show time and again, are often all-too-fallible, or are actually several people, strategy formulators and implementers, working together. Every Kennan needs an Acheson or a Truman; every Louis XIV, a Vauban. But this does not diminish the strategists' crucial importance for reading events before others do and providing timely action. As Bismarck was fond of putting it, "by himself, the individual can create nothing, he can only wait until he listens God's footsteps resounding through events and then spring forward to grasp the hem of his mantle."[46] Strategists figure out the right direction events will take and follow along.

This is not a skill one is born with, but one that requires continuous practice and study, and in particular the study of the efforts of previous famous strategists. One does not always need to reinvent the wheel by struggling to conjure a strategic buzzword out of thin air. The grand strategies of great powers offer models and cautionary tales alike. They are the building blocks for future grand strategies. To this extent, learning from giants, and, in so doing, standing on their shoulders, is never a wasteful pursuit. It gives one a commanding view.

Notes

1 The tables are compiled from data from the Central Intelligence Agency, *CIA World Factbook 2018* at https://www.cia.gov/library/publications/the-world-factbook/ranko rder/2001rank.html; Stockholm Peace Research Institute, "SIPRI Military Expenditure Database," at https://www.sipri.org/databases/milex; and The World Bank, "World Development Indicators," at https://datacatalog.worldbank.org/dataset/world-develop ment-indicators. Data is not currently available for measuring the global innovation index of states prior to recent years.

2 For alternative and additional interpretations of great powers' and regional powers' grand strategies see Balzacq, Dombrovski, and Reich, eds., *Comparative Grand Strategy*; Hitchcock, Leffler, and Legro, eds., *Shaper Nations*; Paul, ed., *Accommodating Rising Power*.

3 Chollet and Goldgeier, *America Between the Wars*; Brands, *From Berlin to Baghdad*; Posen and Ross, "Competing Visions"; Mann, *Rise of the Vulcans*; Daalder and Lindsay, *America Unbound*; Drezner, "Does Obama Have a Grand Strategy"; Daalder and Lindsay, *Empty Throne*; Posen, "Rise of Illiberal Democracy."

4 Buzan, *United States and the Great Powers*, 68–71; Brooks and Wohlforth, *World Out of Balance*.

5 Wohlforth, "Stability of a Unipolar World," 10–22; Posen, "Command of the Commons"; Krauthammer, "The Unipolar Moment," 24; Kennedy, "Eagle Has Landed"; Onea, *US Foreign Policy in the Post-Cold War*.

6 Art, *Grand Strategy for America*, 55–63; Brooks, Ikenberry, and Wohlforth, "Don't Come Home America," 11.

7 Wohlforth, "Stability of a Unipolar World"; Stockholm Peace Research Institute, "Top 10 Military Spenders in 2018," at https://www.sipri.org/research/armament-and-disarm ament/arms-transfers-and-military-spending/military-expenditure; Nina Serafino, *The Department of Defense Role in Foreign Assistance: Background, Major Issues and Options for Congress*, December 2008 available at https://www.fas.org/sgp/crs/natsec/ RL34639.pdf; Kissinger, "Expand NATO Now."

8 Mearsheimer, "Back to the Future"; Nordlinger, *Isolationism Reconfigured*.

9 "Excerpts from the Pentagon's Plan: 'Prevent the Reemergence of a New Rival,'" *New York Times*, March 8, 1992; Patrick Tyler, "US Strategy Plan Calls for Insuring No Rivals Develop," *New York Times*, March 8, 1992; Chollet and Goldgeier, *America Between the Wars*, 43–7; Kagan, and Kristol, "Towards a Neo-Reaganite Foreign Policy."

10 Bush and Scowcroft, *World Transformed*, 400; Alfonsi, *Circle in the Sand*, 109–11; Brands, *From Berlin to Baghdad*, 49–50; Schroeder, "New World Order."

11 Freedman and Karsh, *Gulf Conflict, 1990–1991*.

12 Chollet and Goldgeier, *America Between the Wars*, 130–1; Daalder, *Getting to Dayton*.

13 Although partisans of restraint mention the example of Great Britain's grand strategy, as Chapter Five shows, Great Britain's strategy was actually different because aloof-ness in the core was pursued concomitantly with the exercise of primacy in the periphery. By contrast, the pursuit of restraint would result in detachment all-around, with the US largely keeping to itself.

14 Supporters of restraint are divided as to whether alliances should be preserved in East Asia—some argue that they should be maintained due to the presence of a powerful China that might become a threat too large to handle for local powers.

15 Posen, *Restraint: A New Foundation for US Grand Strategy*; Mearsheimer, *Tragedy*; Walt, *Taming American Power*; Gholz, Press, and Sapolsky, "Come Home, America"; Layne, *Peace of Illusions*; Layne, "From Preponderance to Offshore Balancing."

16 Brooks and Wohlforth, *America Abroad*; Brooks, Ikenberry, and Wohlforth, "Don't Come Home America"; Brooks and Wohlforth, *World Out of Balance*; Dueck, *Obama Doctrine*; Norrlof, *America's Global Advantage*.

17 Mark Manyin, et al., "Pivot to the Pacific: The Obama Administration's 'Rebalancing' Toward Asia," available at http://www.fas.org/sgp/crs/natsec/R42448.pdf; White House, *National Security Strategy of the United States*, December 2017, at https:// www.whitehouse.gov/wp-content/uploads/2017/12/NSS-Final–12–18–2017-0905.p df; Friedberg, *Contest for Supremacy*.

18 Beckley, "China's Century?"; Brooks and Wohlforth, *America Abroad*, chaps. 2–3; Allison, *Destined for War?*; Subramanian, "The Inevitable Superpower"; Jacques, *When China Rules the World*.

19 Rolland, *China's Eurasian Century*, 91, 119; Economy, *Third Revolution*, 194; Sørensen, "Significance of Xi Jinping's 'Chinese Dream'", 53, 65, 68; "Xi Jinping and the Chinese Dream," *The Economist*, May 4, 2013. It should, however, be noted that China does not have an explicit grand strategy, and that it is reluctant to put forward one. Its project to expand westward, first dubbed the One Belt, One Road strategy (OBOR), was rebranded as the Belt and Road Initiative, to avoid negative connotations of the term strategy. Wang, "China's Search for a Grand Strategy"; Angela Stenzel, ed., *Grand Designs: Does China Have a Grand Strategy* at https://www.ecfr.eu/publicatio ns/summary/grands_designs_does_china_have_a_grand_strategy.

20 Kenneth Lieberthal and Wang Jisi, "Addressing US-China Strategic Distrust," at http: //www.brookings.edu/~/media/research/files/papers/2012/3/30%20us%20china%20 lieberthal/0330_china_lieberthal.pdf; Bader, *Obama and China's Rise*.

21 Wang, *Never Forget National Humiliation*; Gries, *China's New Nationalism*, chap. 3; Fairbank, ed., *Chinese World Order*, 1–14.

22 Yan, "The Rise of China in Chinese Eyes," 34; Ye, *Inside China's Grand Strategy*, 74; Deng, *China's Struggle for Status*, 8–9; Economy, *Third Revolution*, 3–4.

23 Ibid., 186–90; Sørensen, "Significance of Xi Jinping's 'Chinese Dream,'" 59–67; Yan, "From Keeping a Low Profile to Striving for Achievement," 164–9. For how a potential US blockade of China would play out see Mirski, "How a Massive Naval Blockade Could Bring China to Its Knees."

24 State Council Information Office of the Popular Republic of China, *China's Military Strategy, May 2015*, 14–15 at https://jamestown.org/wp-content/uploads/2016/07/C hina%E2%80%99s-Military-Strategy–2015.pdf; Economy, *Third Revolution*, 200–4; Stenzel, ed., *Grand Designs*, 7; Corr, ed., *Great Powers, Grand Strategies*, chaps. 1–2; Fravel, "China's Strategy in the South China Sea," 294–6, 307–10; Cliff, Burles, Chase, Eaton, and Pollpeter, *Entering the Dragon's Lair*, chap. 3.

25 See the BRI map at http://news.xinhuanet.com/english/2016-05/15/c_135360904.htm; for the most comprehensive discussion see Rolland, *China's Eurasian Century*, 45–52, 72–88. The BRI has even more ambitious plans: the creation of a future cyber "road" to reinforce the transport network, and of an "Ice Road," linking countries across the Arctic. Economy, *Third Revolution*, 226–9.

26 Rolland, *China's Eurasian Century*, 113, 111–13.

27 Wang, "Marching Westwards," at https://opinion.huanqiu.com/article/9CaKrnJxoLS; Rolland, *China's Eurasian Century*, 116–17.

28 Ibid., 114–16, 96; Economy, *Third Revolution*, 205–18.

29 Wang identifies four possible risks: China may get stuck in quagmires of its own in Eurasia; some countries in the area are caught up in rivalries, which if badly managed, may turn some of them against China; breaking away from containment may preclude diplomatic openings to the US; and China may succumb to the neo-colonial temptation of resource grabbing. Wang, "Marching Westwards."

30 China has met setbacks as a number of countries in the BRI have started suspecting it of "debt-trapping," essentially providing money it knows that the country in question cannot repay in order to gain control over key infrastructure. A further complication has been the accusation that China is using part of this infrastructure for espionage, notably in the building by Huawei of the first 5G network. See for instance Dionne Searcey and Jaime Yaya Barry, "One African Nation Put the Brakes on Chinese Debt. But Not for Long," *New York Times*, November 23, 2018; Maria Habib, "How China Got Sri Lanka to Cough Up a Port," *New York Times*, June 25 2018; David Sanger et al., "In 5G Race with China, US Pushes Allies to Fight Huawei," *New York Times*, January 26, 2019.

31 Vladimir Putin, "Annual Address to the Federal Assembly of the Russian Federation, April 25, 2005," at http://en.kremlin.ru/events/president/transcripts/22931; "Putin: If He Could, He'd Try to Prevent 1991 USSR Collapse, March 2, 2018," at https://ap

news.com/d36b368c6ad44bb2b8e883fc8d800514/Putin:-If-he-could,-he'd-try-to-prevent–1991-USSR-collapse.

32 Shevtsova, *Yeltsin's Russia*; Aslund, *Russia's Capitalist Revolution*; Hoffman, *Oligarchs*; Evangelista, *Chechen Wars*; Goldgeier and McFaul, *Power and Purpose*; Kennan, "Fateful Error"; Daalder and O' Hanlon, *Winning Ugly*.

33 Vladimir Putin, "Russia at the Turn of the Millennium, December 30, 1999," at https://pages.uoregon.edu/kimball/Putin.htm. On Putin's political ascension see Hill and Gaddy, *Mr. Putin*.

34 Shevtsova, *Putin's Russia*; Roxburgh, *Strongman*; Jonas Grätz, "Russia's Military Reforms: Progress and Hurdles," at https://css.ethz.ch/content/dam/ethz/special-inter est/gess/cis/center-for-securities-studies/pdfs/CSSAnalyse152-EN.pdf.

35 Tsygangov, *Russia's Foreign Policy*, chaps. 3–6; Trenin, "Russia's Sphere of Interest"; Trenin, *Should We Fear Russia?*

36 Tsygangov, *Russia's Foreign Policy*, 175–7; Vladimir Putin, "Address by the President of the Russian Federation, March 18, 2014," at http://en.kremlin.ru/events/president/ news/20603. In another context, Putin argued that the colored revolutions represented "a lesson and a warning" for Russia not to allow anything similar occurring its own territory. "Putin Says that Russia Must Prevent 'Colour Revolution,' November 20, 2014," at https://www.reuters.com/article/us-russia-putin-security-idUSKCN0J41J6 20141120.

37 Vladimir Putin, "Speech and the Following Discussion at the Munich Conference of Security Policy, February 10, 2007," at http://en.kremlin.ru/events/president/transcr ipts/24034; Cornell and Starr, *Guns of August 2008*.

38 Charap and Colton, *Everyone Loses*; Freedman, *Ukraine*; Sakwa, *Frontline Ukraine*.

39 Steven Woehrel, "Russian Energy Policy Toward Neighboring Countries," at https://fas .org/sgp/crs/row/RL34261.pdf; Steven Erlanger, "Pipelines from Russia Cross Political Lines," *New York Times*, October 7, 2019; Stanley Reed, "OPEC and Russia Seek to Support the Price of Oil," *New York Times*, June 29, 2019.

40 Michael Gordon, "US Says Russia Tested Cruise Missile, Violating Treaty," *New York Times*, July 28, 2014; Steven Simon, "Hypersonic Missiles Are a Game Changer," *New York Times*, January 2, 2020.

41 Dmitry Adamsky, "Putin's Syria Strategy: Russian Airstrikes and What Comes Next," at https://www.foreignaffairs.com/articles/syria/2015–10-01/putins-syria-strategy; David Kirkpatrick, "Russian Snipers, Missiles, and Warplanes Try to Tilt Libyan War," *New York Times*, November 5, 2019; Eric Schmitt and Thomas Gibbons-Neff, "Russia Exerts Growing Influence in Africa, Worrying Many in the West," *New York Times*, January 28, 1990.

42 Scott Shane and Mark Mazzetti, "The Plot to Subvert and Election: Unravelling the Russia Story So Far," *New York Times*, September 20, 2018; Heidi Reisinger and Alexander Golts, "Russia's Hybrid Warfare: Waging War Below the Radar of Traditional Collective Defence," at http://www.ndc.nato.int/news/news.php?icode =732.

43 Michael Shear and Peter Baker, "Obama Answers Critics, Dismissing Russia as a 'Regional Power,'" *New York Times*, March 25, 2014.

44 Thierry Balzacq, Peter Dombrowski, and Simon Reich, "Is Grand Strategy a Research Program? A Review Essay," *Security Studies*, published online October 2018 at https:/ /www.tandfonline.com/doi/full/10.1080/09636412.2018.1508631, 21–3.

45 Jervis, "Do Leaders Matter," 154, 158–60; Byman and Pollack, "Let Us Now Praise Great Men"; Saunders, *Leaders at War*.

46 Freedman, *Strategy*, 238–9; Gray, *Modern Strategy*, 23–43; Pflanze, *Bismarck and the Development of Germany*, 80, 83.

Bibliography

Alfonsi, Christian. *Circle in the Sand: Why We Went Back to Iraq*. New York: Doubleday, 2006.

Allison, Graham. *Destined for War? Can America and China Escape Thucydides' Trap*. Boston: Houghton Mifflin, 2017.

Art, Robert. *A Grand Strategy for America*. Ithaca: Cornell University Press, 2003.

Aslund, Anders. *Russia's Capitalist Revolution: Why Market Reform Succeeded and Democracy Failed*. Washington: Peterson Institute for International Economics, 2007.

Bader, Jeffrey. *Obama and China's Rise: An Insider Account of America's Asia Strategy*. Washington: Brookings Institution, 2012.

Balzacq, Thierry, Peter Dombrovski, and Simon Reich, eds. *Comparative Grand Strategy: A Framework and Cases*. Oxford: Oxford University Press, 2019.

Beckley, Michael. "China's Century? Why America's Edge Will Endure." *International Security* 36 (Winter 2011): 41–78.

Brands, Hal. *From Berlin to Baghdad: America's Search for Purpose in the Post-Cold War World*. Lexington: University Press of Kentucky, 2008.

Brooks, Stephen, G John Ikenberry, and William Wohlforth. "Don't Come Home America: The Case Against Retrenchment." *International Security* 37 (Winter 2012): 7–51.

Brooks, Stephen and William Wohlforth. *America Abroad: The United States' Global Role in the Twenty-First Century*. New York: Oxford University Press, 2016.

Brooks, Stephen and William Wohlforth. *World Out of Balance: International Relations and the Challenge of American Primacy*. Princeton: Princeton University Press, 2008.

Bush, George HW and Brent Scowcroft. *A World Transformed*. New York: Knopf, 1998.

Buzan, Barry. *The United States and the Great Powers: World Politics in the Twenty-First Century*. New York: Polity, 2004.

Byman, Daniel and Kenneth Pollack. "Let Us Now Praise Great Men: Bringing the Statesman Back In." *International Security* 25 (Spring 2001): 107–46.

Charap, Samuel and Timothy Colton. *Everyone Loses: The Ukraine Crisis and the Ruinous Contest for Post-Soviet Eurasia*. New York: Routledge, 2017.

Chollet, Derek and James Goldgeier. *America Between the Wars: From 11/9 to 9/11*. New York: Public Affairs, 2008.

Cliff, Roger, Mark Burles, Michael Chase, Derek Eaton, and Kevin Pollpeter. *Entering the Dragon's Lair: Chinese Anti-Access Strategies and Their Implications for the United States*. Santa Monica: Rand, 2007.

Cornell, Svante and Frederick Starr. *The Guns of August 2008: Russia's War in Georgia*. Armonk: M. E. Sharpe, 2009.

Corr, Anders, ed. *Great Powers, Grand Strategies: The New Game in the South China Sea*. Annapolis: Naval Institute Press, 2018.

Daalder, Ivo. *Getting to Dayton: The Making of America's Bosnia Policy*. Washington: Brookings Institution, 2000.

Daalder, Ivo and James Lindsay. *America Unbound: The Bush Revolution in Foreign Policy*. Washington: Brookings Institution Press, 2003.

Daalder, Ivo and James Lindsay. *The Empty Throne: America's Abdication of Global Leadership*. New York: Public Affairs, 2018.

Daalder, Ivo and Michael O'Hanlon. *Winning Ugly: NATO's War to Save Kosovo*. Washington: Brookings Institution Press, 2000.

Deng, Yong. *China's Struggle for Status: The Realignment of International Relations.* Cambridge: Cambridge University Press, 2008.

Drezner, Daniel. "Does Obama Have a Grand Strategy: Why We Need Doctrines in Uncertain Times." *Foreign Affairs* 90 (July 2011): 57–68.

Dueck, Colin. *The Obama Doctrine: American Grand Strategy Today.* New York: Oxford University Press, 2015.

Economy, Elizabeth. *The Third Revolution: Xi Jinping and the New Chinese State.* New York: Oxford University Press, 2018.

Evangelista, Matthew. *The Chechen Wars: Will Russia Go the Way of the Soviet Union?* Washington: Brookings Institution, 2002.

Fairbank, John King, ed. *The Chinese World Order: Traditional China's Foreign Relations.* Harvard: Cambridge University Press, 1968.

Fravel, Taylor. "China's Strategy in the South China Sea." *Contemporary Southeast Asia* 33, no. 3 (2011): 292–319.

Freedman, Lawrence. *Strategy.* Oxford: Oxford University Press, 2013.

Freedman, Lawrence. *Ukraine and the Art of Strategy.* New York: Oxford University Press, 2019.

Freedman, Lawrence and Ephraim Karsh. *The Gulf Conflict, 1990–1991: Diplomacy and War in the New World Order.* London: Faber and Faber, 1993.

Friedberg, Aaron. *A Contest for Supremacy: China, America, and the Struggle for Mastery in Asia.* New York: W. W. Norton, 2011.

Gholz, Eugene, Darryl Press, and Harvey Sapolsky. "Come Home America: The Strategy of Restraint in the Face of Temptation." *International Security* 21 (Spring 1997): 5–48.

Goldgeier, James and Michael McFaul. *Power and Purpose: US Policy Toward Russia after the Cold War.* Washington: Brookings Institution Press, 2003.

Gray, Colin. *Modern Strategy.* Oxford: Oxford University Press, 1999.

Gries, Peter. *China's New Nationalism: Pride, Politics, and Diplomacy.* Berkeley: University of California Press, 2004.

Hill, Fiona and Clifford Gaddy. *Mr. Putin: Operative in the Kremlin.* Washington: Brookings Institution Press, 2013.

Hitchcock, William, Melvin Leffler, and James Legro, eds. *Shaper Nations: Strategies for a Changing World.* Cambridge: Harvard University Press, 2016.

Hoffman, David. *The Oligarchs: Wealth and Power in the New Russia.* New York: Public Affairs, 2002.

Jacques, Martin. *When China Rules the World: The Rise of the Middle Kingdom and the End of the Western World.* New York: Penguin, 2009.

Jervis, Robert. "Do Leaders Matter and How Would We Know?" *Security Studies* 22 (2013): 153–79.

Kagan, Robert and William Kristol. "Towards a Neo-Reaganite Foreign Policy." *Foreign Affairs* 75(July/August 1996): 18–32.

Kennan, George. "A Fateful Error." *New York Times*, February 5, 1997.

Kennedy, Paul. "The Eagle Has Landed." *Financial Times*, February 2, 2002.

Kissinger, Henry. "Expand NATO Now." *Washington Post*, December 19, 1994.

Krauthammer, Charles. "The Unipolar Moment." *Foreign Affairs* 70 (Winter 1990): 23–33.

Layne, Christopher. "From Preponderance to Offshore Balancing: America's Future Grand Strategy." *International Security* 22 (Summer 1997): 86–124.

Layne, Christopher. *The Peace of Illusions: American Grand Strategy from 1940 to the Present.* Ithaca: Cornell University Press, 2006.

Mann, James. *Rise of the Vulcans: The History of Bush's War Cabinet*. New York: Viking Books, 2004.

Mearsheimer, John. "Back to the Future: Instability in Europe After the Cold War." *International Security* 15 (Summer 1990): 5–56.

Mearsheimer, John. *Tragedy of Great Power Politics*. New York: W. W. Norton, 2001.

Mirski, Sean. "How a Massive Naval Blockade Could Bring China to Its Knees in a War." *National Interest*, April 6, 2019.

Nordlinger, Eric. *Isolationism Reconfigured: American Foreign Policy for a New Century*. Princeton: Princeton University Press, 1995.

Norrlof, Carla. *America's Global Advantage: US Hegemony and International Cooperation*. New York: Cambridge University Press, 2010.

Onea, Tudor. *US Foreign Policy in the Post-Cold War: Restraint Versus Assertiveness from George H. W. Bush to Barack Obama*. New York: Palgrave Macmillan, 2013.

Paul, TV, ed. *Accommodating Rising Powers: Past, Present, and Future*. Cambridge: Cambridge University Press, 2016.

Pflanze, Otto. *Bismarck and the Development of Germany: The Period of Unification, 1815–1871*, vol. 1. Princeton: Princeton University Press, 1990.

Posen, Barry. "Command of the Commons: The Military Foundation of U.S. Hegemony." *International Security* 28 (Summer 2003): 5–46.

Posen, Barry. *Restraint: A New Foundation for US Grand Strategy*. Ithaca: Cornell University Press, 2014.

Posen, Barry. "The Rise of Illiberal Democracy: Trump's Surprising Grand Strategy." *Foreign Affairs* 97 (March/April 2018): 20–7.

Posen, Barry and Andrew Ross. "Competing Visions for US Grand Strategy." *International Security* 21 (Winter 1996/7): 5–53.

Rolland, Nadège. *China's Eurasian Century? Political and Strategic Implications of the Belt and Road Initiative*. Seattle: National Bureau of Asian Research, 2017.

Roxburgh, Angus. *The Strongman: Vladimir Putin and the Struggle for Russia*. London: I. B. Tauris, 2012.

Sakwa, Richard. *Frontline Ukraine: Crisis in the Borderlands*. London: I. B. Tauris, 2015.

Saunders, Elizabeth. *Leaders at War: How Presidents Shape Military Interventions*. Ithaca: Cornell University Press, 2011.

Schroeder, Paul. "The New World Order: A Historical Perspective." *Washington Quarterly* 17 (Spring 1994): 25–44.

Shevtsova, Lilia. *Putin's Russia*. Washington: Brookings Institution Press, 2005.

Shevtsova, Lilia. *Yeltsin's Russia: Myths and Reality*. Washington: Brookings Institution, 1999.

Sørensen, Camilla. "The Significance of Xi Jinping's 'Chinese Dream' for Chinese Foreign Policy: From 'Tao Guang Yang Hui' to 'Fen Fa You Wei'." *Journal of Chinese International Relations* 3, no. 1 (2015): 53–73.

Subramanian, Arvind. "The Inevitable Superpower: Why China's Rise Is a Sure Thing." *Foreign Affairs* 90 (September 2011): 66–78.

Trenin, Dmitri. "Russia's Sphere of Interest, Not Influence." *Washington Quarterly* 32, no. 4 (2009): 3–22.

Trenin, Dmitri. *Should We Fear Russia?*. Cambridge: Polity Press, 2016.

Tsygangov, Andrei. *Russia's Foreign Policy: Change and Continuity in National Identity*, 3rd edn. Lanham: Rowman & Littlefield, 2013.

Walt, Stephen. *Taming American Power: The Global Response to US Primacy*. New York: W. W. Norton & Company, 2005.

Wang, Jisi. "China's Search for a Grand Strategy: A Rising Great Power Finds Its Way." *Foreign Affairs* 90, no. 2 (2011): 68–79.

Wang, Jisi. "Marching Westwards: The Rebalancing of China's Geostrategy." *International and Strategic Studies Report*, Centre for International and Strategic Studies Beijing University 73, October 7, 2012.

Wang, Zheng. *Never Forget National Humiliation: Historical Memory in Chinese Politics and Foreign Relations*. New York: Columbia University Press, 2014.

Wohlforth, William. "The Stability of a Unipolar World." *International Security* 24 (Summer 1999): 5–41.

Yan, Xuetong. "From Keeping a Low Profile to Striving for Achievement." *Chinese Journal of International Politics* 7, no. 2 (2014): 153–184.

Yan, Xuetong. "The Rise of China in Chinese Eyes." *Journal of Contemporary China* 10, no. 26 (2001): 33–9.

Ye, Zicheng. *Inside China's Grand Strategy: The Perspective from the People's Republic*. Lexington: University Press of Kentucky, 2011.

Index